T0275973

SpringerBriefs in Intelligent Systems

Artificial Intelligence, Multiagent Systems,
and Cognitive Robotics

Series editors

Gerhard Weiss, Maastricht, The Netherlands
Karl Tuyls, Liverpool, UK

More information about this series at http://www.springer.com/series/11845

Neng-Fa Zhou · Håkan Kjellerstrand
Jonathan Fruhman

Constraint Solving and Planning with Picat

 Springer

Neng-Fa Zhou
Department of Computer and Information
   Science
Brooklyn College
Brooklyn, NY
USA

Jonathan Fruhman
Independent Researcher
New York, NY
USA

Håkan Kjellerstrand
hakank.org
Malmö
Sweden

ISSN 2196-548X   ISSN 2196-5498   (electronic)
SpringerBriefs in Intelligent Systems
Artificial Intelligence, Multiagent Systems, and Cognitive Robotics
ISBN 978-3-319-25881-2   ISBN 978-3-319-25883-6   (eBook)
DOI 10.1007/978-3-319-25883-6

Library of Congress Control Number: 2015954588

Springer Cham Heidelberg New York Dordrecht London

Springer International Publishing AG Switzerland is part of Springer Science+Business Media
(www.springer.com)

Foreword

Logic programming languages entered the scene of computer science in the early 1970s as the answer to the need for paradigms capable of representing and reasoning about different kinds of knowledge. The big picture that was pursued by logic programming researchers was to create a black-box system that is able to transform declarative and logic-based specifications into actionable problem solutions ("the holy grail of programming," as some authors say). The history of logic programming witnessed periods of genuine enthusiasm and leaps of knowledge, and also witnessed periods of stagnation. The initial proposal of the language Prolog by Robert Kowalski came with the drawback of the inefficiency of its first implementation of the SLD resolution procedure. As a consequence, the AI community, looking for fast implementations of planning algorithms, moved towards other directions—from the (functionally inspired) PDDL modeling language, which is the de facto standard formalism for planning problem descriptions, to the use of traditional imperative languages.

In the early 1980s, results from unification theory and the implementation of the Warren Abstract Machine by D.H.D. Warren sensibly improved the efficiency of the Prolog implementations, enabling the compilation of programs into efficiently executable bytecode. In those same years, the AI problem-solving community developed the notion of constraint propagation and constraint-based search; these declarative formalisms required a host language in order to be effectively exploited. Researchers in the US (e.g., Joxan Jaffar, Jan-Louis Lassez, Michael Maher, and Peter Stuckey) and in Europe (e.g., Pascal Van Hentenryck), on the wave of the Fifth Generation project, offered effective frameworks and implementations to enable the parametric extension of the Prolog language into a constraint-based framework. Constraint logic programming is a class of logic programming languages whose modeling capabilities are based on the external constraint solver chosen by the programmer (e.g., a solver on finite domains, on rational numbers, or on sets). This extension of Prolog allowed the efficient resolution of large classes of practical problems, and constituted a common research path for the logic programming and constraint programming communities.

However, logic programming and constraint logic programming continued to fall short of expectations in the context of knowledge representation and reasoning. The original declarative semantics that were developed for Prolog and preserved by constraint logic programming were struggling to meet the needs of non-monotonic reasoning. An elegant and effective solution to this problem arose from the work of Michael Gelfond and Vladimir Lifschitz, which was embodied by the stable model semantics for logic programming with negation as failure. It took almost 10 years for researchers to understand how to effectively embed such semantics in a Prolog-style language, leading to a novel paradigm, commonly referred to as answer set programming (ASP). Planning problems are naturally encoded in ASP, and can also be directly encoded as a SAT formula, exploiting the progress in this area. In this millennium, the idea of conflict-driven learning has been introduced in both ASP solvers and SAT solvers, making these systems highly effective for solving large and complex problems. In parallel with the evolution of ASP, the notion of tabling/memoization was introduced by D.S. Warren, and was implemented in Prolog systems (including B-Prolog by Neng-Fa Zhou). Tabling allows Prolog systems to avoid the recomputation of sub-goals, and can also be used for supporting dynamic programming-style optimizations. A good use of tabling has been proven effective for speeding up the search, particularly in solving planning problems.

The language Picat, the subject of this book, is the culminating event of these developments. The language is as declarative as Prolog, but it is more convenient than Prolog in many aspects. It supports the encoding of problems by using constraints, and it enables the search for solutions through the use of constraint and SAT solvers. Picat provides the use of tabling with all of its features. In particular, Picat allows the encoding of planning problems in a programming style that is similar to PDDL, and facilitates their fast resolution thanks to tabling.

I do not know if Picat is already "the holy grail," but surely a generation of students/programmers can benefit a lot from this language. This book provides the perfect compendium to enter the fascinating universe of Picat.

Udine   Agostino Dovier
September 2015

Preface

Many complex systems, ranging from social, industrial, economics, financial, educational, to military, require that we obtain high-quality solutions to combinatorial problems. Linear programming and its extensions developed in operations research once formed the primary paradigm for solving combinatorial problems. During the last three decades, a plethora of paradigms have been developed for combinatorial problems, including constraint programming (CP), propositional satisfiability testing (SAT), satisfiability modulo theories (SMT), answer set programming (ASP), tabled logic programming, and heuristic search planners.

Picat is a new logic-based multi-paradigm programming language that integrates logic programming, functional programming, dynamic programming with tabling, and scripting. Picat provides facilities for solving combinatorial search problems, including solver modules that are based on CP, SAT, and MIP (mixed integer programming), and a module for planning that is implemented by the use of tabling. Chapter 1 gives an overview of the Picat language and system.

This book presents Picat as a modeling and solving language for two important classes of combinatorial problems: constraint satisfaction and planning. The constraint satisfaction problem (CSP) is a basic class of combinatorial problems. A CSP consists of a set of variables, each of which is defined over a domain, a set of constraints among the variables, and, optionally, an objective function. A solution to a CSP is a valuation of the variables that satisfies all the constraints and optimizes the objective function, if it exists. Chapters 2 and 3 are devoted to constraint modeling. Chapter 2 introduces the basic built-in constraints in the common API of the three solver modules (cp, sat, and mip), and gives several simple example models that use these constraints. Chapter 3 describes the more sophisticated constraints in the API and gives "real-world" example models for scheduling, resource allocation, and design.

Planning is an important task for building many systems, such as autonomous robots, industrial design software, system security, and military operational systems. Planning is also closely related to model checking. Given an initial state, a set of goal states, and a set of possible actions, the classic planning problem is to find a plan that transforms the initial state to the goal state. Chapters 5 and 6 are devoted to planning. Chapter 5 introduces depth-unbounded search predicates in the planner module, and gives models for several planning puzzles. Chapter 6 introduces depth-bounded search predicates in the planner module, and shows how domain knowledge and heuristics can be incorporated into planning models to improve the performance. Because planning is solved as a dynamic programming problem with tabling in Picat, a separate chapter (Chap. 4) is included to introduce the tabling feature of Picat.

A combinatorial problem can normally be modeled in different ways and solved by different solvers. The book is wrapped up with a chapter (Chap. 7) that gives several models for solving the Traveling Salesman Problem with different solvers, including CP, SAT, MIP, and tabled planning.

This book is useful for students, including undergraduate-level students, researchers, and practitioners, to learn the modeling techniques for Picat. No pre-requisite knowledge about Picat is required, although familiarity with logic or functional programming is a plus. Chapter 1 gives an overview of the data types, built-ins, and language constructs that are needed for the later chapters. For readers who are familiar with Prolog, Haskell, or Python, this chapter also compares Picat with these three languages.

This book does not cover the implementation of Picat. The bibliographical note at the end of each chapter aims to meet the reader's curiosity about how the presented tools are built. Each chapter ends with a set of exercises, which is intended for the reader to practice the presented modeling techniques. The code examples used in the book are available at: http://picat-lang.org/picatbook2015. html.

Neng-Fa Zhou would like to thank his other co-authors for their collaboration on some of the important ideas that underpin the Picat implementation: Roman Barták, Agostino Dovier, Christian Theil Have, Taisuke Sato, and Yi-Dong Shen; his recent students at CUNY who worked on application projects using Picat: Mike Bionchik and Lei Chen; his colleagues at CUNY who have been involved in some way in the Picat project: David Arnow, James Cox, Danny Kopec, and Subash Shankar. The authors would like to thank Ronan Nugent, the Springer editor, for his helpful comments and advice, and the following people who gave us permission to use their examples: Brian Dean, Christopher Jefferson, Tony Hürlimann, and Peter James Stuckey.

Brooklyn, NY, USA   Neng-Fa Zhou
Malmö, Sweden   Håkan Kjellerstrand
New York, NY, USA   Jonathan Fruhman
August 2015

Contents

Chapter 1
An Overview of Picat

Abstract Picat is a general-purpose programming language that includes features from multiple programming paradigms. By combining imperative programming's control flow features with more advanced features from logic programming, functional programming, and scripting, Picat provides users a wide array of tools for efficiently creating programs to solve problems. A comparison with other programming languages shows how Picat programs can be more compact, more intuitive, and easier to read.

1.1 Introduction

Picat[1] is a programming language that incorporates features from multiple programming paradigms. The purpose of Picat is to bridge the gap between imperative languages, which tell the computer how things should be done, and declarative languages, which tell the computer what to do, without detailing how it should be done. The inclusion of features from different programming families allows users to create short, powerful, programs for a wide variety of applications.

For example, the following short code implements the *quicksort* algorithm. This code will be explained in Sect. 1.4.3. For now, note that the algorithm can be implemented in just a few lines of Picat code.

```
qsort([]) = [].
qsort([H|T]) = L =>
    L = qsort([E : E in T, E=<H]) ++
        [H] ++
        qsort([E : E in T, E>H]).
```

Logic programming languages, such as Prolog, use a series of facts and rules in order to define relationships. Logic programs find values that satisfy these relationships. Picat uses rule-based *predicates* in order to define relations, as described in

---

[1]Picat is freely available at http://picat-lang.org.

© The Author(s) 2015

N.-F. Zhou et al., *Constraint Solving and Planning with Picat*,
SpringerBriefs in Intelligent Systems, DOI 10.1007/978-3-319-25883-6_1

Sect. 1.4.1. Logic programs can *backtrack* in order to return multiple sets of values that satisfy relationships. Backtracking is discussed in Sect. 1.4.1. Picat's logic programming features also include a *unification operator* that binds variables to values. The unification operator is detailed in Sect. 1.2.4.

The following is an example of backtracking:

```
Picat> member(X,[1,2,3])
X = 1;
X = 2;
X = 3;
no
```

This example demonstrates how Picat's member predicate backtracks. The first line of this example represents input at the Picat interpreter, and the other lines represent output. After Picat displays X = 1, the user types a semicolon (;), causing the member predicate to backtrack, and to output X = 2. After this process repeats, finding X = 3, no more solutions remain, and the system prints no.

Imperative programming languages, such as C and Java, focus on a program's implementation details. Picat incorporates imperative programming aspects that control the flow of a program, such as if-then-else statements and foreach loops. *Control flow* is discussed in Sect. 1.3, with Sect. 1.3.2 focusing on if-then-else statements, and Sect. 1.3.4 focusing on foreach loops. Furthermore, imperative programming languages have assignment statements that update a variable's value. Picat has an *assignment operator*, described in Sect. 1.3.3, that simulates traditional imperative assignment and re-assignment.

Functional programming languages, such as Haskell, treat a program as a series of recursively-defined functions. *Pattern-matching* is used to define functions. Section 1.4 shows the pattern-matching rules that Picat uses to define predicates and functions. Section 1.4 also demonstrates how predicates and functions can be defined recursively, in terms of themselves. Another feature of functional languages is the ability to easily manipulate lists of values. Picat lists are described in Sect. 1.2.3.

Scripting languages, such as Python and Perl, are dynamic languages that provide language constructs for concisely describing computations and for gluing a series of tasks together. Scripting languages use an *interpreter* that can automate a series of tasks. Picat includes a command-line interpreter that enables users to evaluate the results of one or more statements.

1.1.1 Running Picat

The Picat interpreter is started with the OS command picat. Although Picat can interpret a series of statements, it can also compile and run programs that are saved in files. The built-in predicate cl (*FileName*) compiles and loads the source file named

FileName.pi within the Picat interpreter. This allows users to run a program's predicates and functions. If the program has a predicate called main, the command picat *FileName Arg*$_1$ ... *Arg*$_n$ can be used to run the program as a standalone.

Note that Picat has many more features than are covered in this chapter. For a comprehensive coverage of Picat's features, see the Picat User's Guide.[2]

1.2 Data Types and Operators

Picat is a dynamically-typed language, in which type checking occurs at runtime. Table 1.1 lists all of the operators that Picat provides. Picat also includes a % symbol, which begins single-line comments, and the /* and */ symbols, which surround multi-line comments.

1.2.1 Terms, Variables, and Values

Picat is a dynamically-typed language, in which type checking occurs at runtime. A *term* is a variable or a value. A variable gets a type once it is bound to a value. Variables in Picat, like variables in mathematics, are value holders. Unlike variables in imperative languages, Picat variables are not symbolic addresses of memory locations. A variable is said to be *free* if it does not hold any value. A variable is *instantiated* when it is bound to a value. Picat variables are *single-assignment*, which means that after a variable is instantiated to a value, the variable will have the same identity as the value. After execution backtracks over a point where a binding took place, the value that was assigned to a variable will be dropped, and the variable will be turned back into a free variable.

A variable name is an identifier that begins with a capital letter or the underscore. For example, the following are valid variable names:

```
X1    _    _ab
```

The name _ is used for *anonymous variables*.

A term is *ground* if the term does not contain any variables. The built-in predicate ground(*Term*) tests whether a term is ground.

A value in Picat can be *primitive* or *compound*.

---

[2]http://picat-lang.org/download/picat_guide.pdf.

Table 1.1 Operators in Picat

Precedence	Operators
Highest	., @
	** (right-associative)
	unary +, unary -, ~
	*, /, //, /<, />, div, mod, rem
	binary +, binary -
	>>, <<
	/\
	^
	\/
	..
	++ (right-associative)
	=, !=, :=, ==, !==, <, =<, <=, >, >=, ::, in, notin
	#=, #!=, #<, #=<, #<=, #>, #>=, @<, @=<, @<=, @>, @>=
	#~
	#/\
	#^
	#\/
	#=> (right-associative)
	#<=>
	not, once
	, (right-associative), && (right-associative)
Lowest	; (right-associative), || (right-associative)

1.2.2 Primitive Values

A primitive value can be an *atom*, an *integer*, or a *real number*.

An *atom* is a symbolic constant. An atom name can be either unquoted or quoted. An unquoted name is an identifier that begins with a lower-case letter, followed by an optional string of letters, digits, and underscores. A quoted name is a single-quoted sequence of arbitrary characters. A character can be represented as a single-character atom.

A number can be an *integer* or a *real number*. An integer can be a decimal numeral, a binary numeral, an octal numeral, or a hexadecimal numeral. A real number consists of an optional integer part, an optional decimal fraction preceded

Table 1.2 Arithmetic operators

$X ** Y$	Power
$+X$	Same as X
$-X$	Sign reversal
$\sim X$	Bitwise complement
$X * Y$	Multiplication
X / Y	Division
$X // Y$	Integer division, truncated
$X /> Y$	Integer division (ceiling(X / Y))
$X /< Y$	Integer division (floor(X / Y))
X div Y	Integer division, floored
X mod Y	Modulo, same as X - floor(X div Y) * Y
X rem Y	Remainder (X - (X // Y) * Y)
$X + Y$	Addition
$X - Y$	Subtraction
$X \gg Y$	Right shift
$X \ll Y$	Left shift
$X /\backslash Y$	Bitwise and
$X \char`^ Y$	Bitwise xor
$X \backslash/ Y$	Bitwise or
From .. Step .. To	A range (list) of numbers with a step
From .. To	A range (list) of numbers with step 1

by a decimal point, and an optional exponent. Table 1.2 gives the meaning of each of the numeric operators in Picat, from highest precedence to lowest precedence.

1.2.3 Compound Terms

A compound term can be a *list*, a *string*, or a *structure*. Picat defines two types of special structures: *arrays* and *maps*.

A *list* takes the form $[t_1, \ldots, t_n]$, where each t_i ($1 \leq i \leq n$) is a term. The list is represented internally as a singly-linked list. The index notation $L[I]$ is a special function that returns the Ith component of list L, with $L[1]$ referring to the first element of L. In order to represent a multi-dimensional list, an index notation can take multiple subscripts. For example, the expression `L[1,2]` is the same as `T[2]`, where `T` is a temporary variable that references the component that is returned by `L[1]`. The bar symbol, `|`, is a separator that separates the list's first element (called the *head*) from the rest of the list (called the *tail*). The *cons* notation $[H|T]$ can occur in a pattern or in an expression. When it occurs in a pattern, it matches any list in which H matches the head and T matches the tail. When it occurs in an expression, it builds a list from H and T. Note that the pattern $[E_1, E_2|T]$ is the same as $[E_1|[E_2|T]]$. Also note that T could be an empty list.

Picat provides a number of built-ins for lists. The new_list(*N*) function creates a new list that has *N* free variable arguments. The ++ operator concatenates two lists. The length(*List*) function returns the number of elements that are in *List*. This function can also be represented in *dot notation*[3] as *List*. length. Note that the length of a list is not stored in memory; instead, it is recomputed each time that the length function is called, which takes linear time in terms of the list's size.

Picat also includes *list comprehensions*. A list comprehension is a special functional notation for creating lists without specifying the individual values that the list stores. List comprehensions have the following format:

$$[T : E_1 \text{ in } D_1, Cond_1, \ldots, E_n \text{ in } D_n, Cond_n]$$

T is an expression. Each E_i is an *iterating pattern*. Each D_i is an expression that gives a *compound value*. Each $Cond_i$ is an optional *condition* on iterators E_1 through E_i.

The expression E_i in D_i is called an *iterator*, which can take multiple forms. The first form,

> *Value* in *Start* . . *End*

uses the *range, Start* . . *End*, which generates a list that contains every integer from *Start* through *End*, including *Start* and *End*. Another form,

> *Value* in *Start* . . *Step* . . *End*

uses the range *Start* . . *Step* . . *End*. In this case, *Start*, *Step*, and *End* can be integers or real numbers, and the values in the generated list will start at *Start*, and will repeatedly increment by *Step* until *End* is reached or exceeded. Note that increment can be negative. A third form is

> *Element* in *CompoundTerm*

This iterator loops through each *Element* in the specified *CompoundTerm*.

Examples: Lists

```
Picat> L = new_list(5)
L = [_80bc,_80c4,_80cc,_80d4,_80dc]

Picat> L = [[1,2,3],[4,5,6]], Val = L[2,1]
L = [[1,2,3],[4,5,6]]
Val = 4

Picat> L = [1,2,3] ++ [4,5,6]
L = [1,2,3,4,5,6]

Picat> [H|T] = [a,b,c]
H = a
T = [b,c]
```

---

[3] In general, all functions can be called by using dot notation.

```
Picat> L = [I : I in 0 .. 2 .. 10]
L = [0,2,4,6,8,10]

Picat> L = [I : I in 0 .. 10, I mod 2 == 0]
L = [0,2,4,6,8,10]

Picat> L = [I : I in 1 .. -0.5 .. -1]
L = [1,0.5,0.0,-0.5,-1.0]

Picat> L = [(C, I) : C in [a, b], I in 1 .. 2]
L = [(a,1),(a,2),(b,1),(b,2)]
```

In each of the above examples, the first line represents input at the Picat inter-preter, and the other lines represent output. The first example creates a new list that has five free variables. The second example creates the two-dimensional list [[1,2,3],[4,5,6]], and demonstrates how the index notation accesses ele-ments of the list. The third example uses the ++ operator, which concatenates two lists. The fourth example shows how the separator | separates the head of a list from the tail. This example also demonstrates the unification operator, which is described in Sect. 1.2.4.

The last four examples demonstrate list comprehensions. The first list compre-hension creates a list of the even numbers that are between 0 and 10. The second list comprehension also creates a list of the even numbers that are between 0 and 10; however, instead of looping through every other number, the second list comprehen-sion loops through every number, and tests the numbers with the *condition* I mod 2 == 0. The third list comprehension demonstrates a negative value for *Step*. The last list comprehension creates a list of pairs. Note that, in the last list comprehension, the first iterator iterates through a compound term, while the second iterator iterates through a range.

A *string*, which appears between double quotes, is represented as a list of single-character atoms. For example, the string "hello" is the same as the list [h,e,l,l,o] and the list ['h','e','l','l','o']. Picat's built-ins for lists, including the ++ operator and the length function, also operate on strings.

A *structure* has the form $s(t_1, \ldots, t_n)$, where s stands for a structure name, n is called the *arity* of the structure, and each t_i ($1 \leq i \leq n$) is a *term*. The dollar symbol that precedes the structure is used to distinguish a structure from a function call.

An *array* is a special structure that takes the form $\{t_1, \ldots, t_n\}$, where each t_i ($1 \leq i \leq n$) is a term. Arrays are similar to lists. Arrays can be created with a new_array function, can be concatenated with the ++ operator, can be passed to the length function, and can be defined by *array comprehensions*. Note that the separator | does not operate on arrays, since arrays are not built by conses. Also note that, unlike a list, an array always has its length stored in memory, so the function length(*Array*) always takes constant time.

   A *map* is a *hash-table* that is represented as a special structure that contains a
set of key-value pairs. Maps are created with the built-in functions new_map(*N*)
and new_map(*PairsList*). The parameter *N* is an integer that represents the map's
initial capacity, while the parameter *PairsList* is a list of pairs, where each pair has
the form *Key* = *Val*. Note that another way to create a map is with the function
new_map(*N*, *PairsList*), which passes both an initial capacity and a list of pairs.
The function size(*Map*) returns the number of key-value pairs that are in *Map*. The
predicate put(*Map*, *Key*, *Val*) adds the key-value pair *Key* = *Val* to *Map*. The
function get(*Map*, *Key*) = *Val* returns the *Val* of the key-value pair *Key* = *Val*.
If *Map* does not contain *Key*, then the get function throws an error. The variant
get(*Map*, *Key*, *Default*) = *Val* returns the default value *Default* if *Map* does not
contain *Key*. The predicate has_key(*Map*, *Key*) determines whether *Map* contains
Key.

Examples: Arrays and Maps

```
Picat> A = new_array(5)
A = {_83bc,_83c0,_83c4,_83c8,_83cc}

Picat> A = {{1,2,3},{4,5,6}}, Val = A[2,1]
A = {{1,2,3},{4,5,6}}
Val = 4

Picat> A = {1,2,3} ++ {4,5,6}
A = {1,2,3,4,5,6}

Picat> A = {(C, I) : C in {a, b}, I in 1 .. 2}
A = {(a,1),(a,2),(b,1),(b,2)}

Picat> M = new_map(5)
M = (map)[]

Picat> M = new_map([one=1,two=2]), put(M, three, 3)
M = (map)[two = 2,one = 1,three = 3]

Picat> M = new_map([one=1,two=2]), Val = get(M, one)
M = (map)[two = 2,one = 1]
Val = 1

Picat> M = new_map([one=1]), has_key(M, two)
no

Picat> M = new_map([one=1]), Val = get(M, two, 0)
M = (map)[one = 1]
Val = 0
```

The first example creates a new array that has five free variables. The second example creates the two-dimensional array {{1,2,3},{4,5,6}}, and demonstrates how the index notation accesses elements of the array. The third example uses the ++ operator to concatenate two arrays. The fourth example demonstrates an array comprehension. Note that array comprehensions are similar to list comprehensions, except that array comprehensions use the symbols {} instead of [].

The next five examples show how to create, modify, and access maps. The first map example creates a map that has an initial capacity of 5. The second map example creates a map that initially has two key-value pairs, and then uses the put predicate to add a third key-value pair to the map. The third map example shows how the get function can be used to retrieve a value from a map. The last two map examples show what happens when a map does not contain a key. The map does not contain the key two, so has_key fails, while get/3 returns the default value 0.

1.2.4 Equality Testing and Unification

The equality test $T_1 == T_2$ is true if term T_1 and term T_2 are identical. Two variables are identical if they are aliases. Two primitive values are identical if they have the same type and the same internal representation. Two lists are identical if the heads are identical and the tails are identical. Two structures are identical if their names and lengths are the same and their components are pairwise identical. The inequality test $T_1 !== T_2$ is the same as not $T_1 == T_2$. Note that two terms can be identical even if they are stored in different memory locations.

The *unification* $T_1 = T_2$ is true if term T_1 and term T_2 are already identical, or if they can be made identical by instantiating the variables in the terms. The built-in $T_1 != T_2$ is true if term T_1 and term T_2 are not unifiable.

Given the unification $T_1 = T_2$: (a) If T_1 is a variable, then the system binds it to T_2. (b) If T_2 is a variable, then the system binds it to T_1. (c) If both T_1 and T_2 are atomic values, then the unification succeeds if the values are identical, and fails if the values are not identical. (d) If $T_1 = \$f(A_1, \ldots, A_m)$ and $T_2 = \$g(B_1, \ldots, B_n)$, then the unification fails if variable $f != g$ or arity $m != n$; otherwise, the unification returns the results of $A_1 = B_1, \ldots, A_m = B_n$.

Examples: Unification

```
Picat> X = 1
X = 1

Picat> $f(X,b) = $f(a,Y)
X = a
Y = b

Picat> $f(X,Y) = $f(a,b,c)
```

```
no

Picat> [H|T] = [a,b,c]
H = a
T = [b,c]
```

The first example binds *X* to 1. The second example unifies two structures by binding *X* to a and *Y* to b. The third example fails, because the two structures do not have the same length. The fourth example matches *H* to the head of the list, and matches *T* to the tail of the list.

1.2.5 I/O and Modules

Picat provides a number of built-ins for performing input and output operations. For example, the function read_int() = *Integer* reads an integer from standard input. In order to write to standard output, Picat includes the predicates writeln(*Term*) and println(*Term*). These predicates write *Term* followed by a newline, meaning that the output of the next write will begin on the next line. To omit the newline, use write(*Term*) or print(*Term*). The nl predicate can be used to print a newline.

The difference between the write and print predicates is that the write predicates place quotes around strings and atoms whenever necessary, so that they remain the same when read back by Picat, while the print predicates do not place quotes around strings and atoms. For example, write('X') outputs 'X', which is still an atom when read back by Picat. However, print('X') outputs X, which becomes a variable when read back by Picat.

Picat has library *module* files that include additional built-ins. In order to use these built-ins from within other modules, a Picat file must begin with an *import declaration* of the form import *Name*$_1$, ..., *Name*$_n$, where each *Name*$_i$ is a module name. The four basic modules, which are basic, math, sys, and io, are imported by default. Picat's other modules include cp, sat, and mip, which are used to solve constraint satisfaction problems, as demonstrated in Chaps. 2 and 3. Another of Picat's modules is the planner module, which is shown in Chaps. 5 and 6.

1.3 Control Flow and Goals

1.3.1 Goals

A *goal* is made from predicate calls and statements. Suppose that *P* and *Q* are goals. The *conjunction* of the goals, written as *P*, *Q* or *P* && *Q*, succeeds if both goals *P* and *Q* succeed. The *disjunction* of the goals, written as *P*; *Q* or *P* || *Q*, succeeds if either *P* or *Q* succeeds. The built-in predicate true always succeeds, and the built-in predicates `fail` and `false` always fails. The *negation* of a goal, written as `not` *P*, succeeds if *P* fails. Through the use of backtracking, as described in Sect. 1.4.1, goals can succeed more than once. The goal `once` *P* can never succeed more than once.

1.3.2 If-Then-Else

Another type of goal is the *if-then-else statement*, which is used to perform different operations, depending on whether or not certain conditions are fulfilled. If-then-else statements take the form

```
if Cond₁ then
    Goal₁
elseif Cond₂ then
    Goal₂
    ⋮
elseif Condₙ then
    Goalₙ
else
    Goal_else
end
```

where the `elseif` and `else` clauses are optional. If the `else` clause is missing, then the `else` goal is assumed to be `true`. For the if-then-else statement, Picat finds the first condition $Cond_i$ that is true. If such a condition is found, then the truth value of the if-then-else statement is the same as $Goal_i$. If none of the conditions is true, then the truth value of the if-then-else statement is the same as $Goal_{else}$.

Note that Picat also accepts if-then-else in the form `(If->Then;Else)`, mandating the presence of the `Else` part. This style of if-then-else matches Prolog's style.

Example: If-Then-Else

```
main =>
    Score = 74,

    % Assigns a letter grade based on the score.
    if Score >= 90 then
       Grade = a
    elseif Score >= 80 then
       Grade = b
    elseif Score >= 70 then
       Grade = c
```

```
elseif Score >= 60 then
   Grade = d
else
   Grade = f
end,


% Determines whether Score passes or fails.
(Score >=60 -> println("Pass"); println("Fail")).
```

This example demonstrates if-then-else statements and Prolog-style if-then-else statements. First, a score is assigned. Then, the if-then-else conditions determine the letter grade. Since the score is 74, the goal `Score >= 70` succeeds, and `Grade` is set to `c`. When `Score >= 70` succeeds, the remaining branches of the statement are not evaluated. The last line of the program uses a Prolog-style if-then-else statement to determine whether the score passes or fails.

The following code can also be used to determine whether the score passes or fails:

```
Message = cond(Score >= 60, pass, fail)
```

This code uses Picat's *conditional expression*, which takes the form *Value* = cond(*Condition,Then,Else*). If *Condition* is true, then *Value* is unified with *Then*; otherwise, *Value* is unified with *Else*. In this example, the score is 74, so `Message` receives a value of `pass`.

This example also shows the use of the *comparison* operators. These operators include >, <, >=, =<, <=, ==, and !==. Note that both the =< and the <= operators indicate less-than-or-equals.

Unlike the previous examples, the code in this example is stored in a `main` predicate. User-defined predicates and functions are detailed in Sect. 1.4.

1.3.3 The Assignment Operator

Picat variables are *single-assignment*, meaning that once a variable is bound to a value, the variable cannot be bound again, unless the value is a variable or the value contains variables. In order to simulate imperative language variables, Picat provides the operator : =, which allows *assignments* in rule bodies. An assignment takes the form *LHS* : = *RHS*, where *LHS* is either a variable or an access of a compound value in the form $X [...]$. When *LHS* is an access in the form $X[I]$, the component of X indexed I is updated. This update is undone if execution backtracks over this assignment.

Example: Assignments

```
main => X=0, X:=X+1, X:=X+2, write(X).
```

In order to handle assignments, Picat creates new variables at compile time. In the above example, at compile time, Picat creates a new variable, say X1, to hold the value of X after the assignment X:=X+1. Picat replaces X by X1 on the LHS of the assignment. It also replaces all of the occurrences of X to the right of the assignment by X1. When encountering X1:=X1+2, Picat creates another new variable, say X2, to hold the value of X1 after the assignment, and replaces the remaining occurrences of X1 by X2. When write(X2) is executed, the value held in X2, which is 3, is printed. This means that the compiler rewrites the above example as follows:

```
main => X=0, X1=X+1, X2=X1+2, write(X2).
```

1.3.4 Foreach Loops

Picat provides *foreach loops*, which are imperative programming constructs for performing repetitions. A foreach loop has the form:

$$\text{foreach} \ (E_1 \ \text{in} \ D_1, \ Cond_1, \ \ldots, \ E_n \ \text{in} \ D_n, \ Cond_n)$$
$$\qquad Goal$$
$$\text{end}$$

The form of foreach loops is similar to the form of list comprehensions, as described in Sect. 1.2.3. Each E_i is an iterating pattern. Each D_i is an expression that gives a compound value. Each $Cond_i$ is an optional condition on iterators E_1 through E_i.

Example: Transpose an Array

```
main =>
    % sample matrix
    Matrix = {{1,2,3},{4,5,6}},

    N = length(Matrix),
    M = length(Matrix[1]),
    Transposed = new_array(M,N),

    % transposes the matrix
    foreach (I in 1..N, J in 1..M)
       Transposed[J,I] = Matrix[I,J]
    end,

    % prints the transposed matrix
```

```
foreach (Row in Transposed)
   println(Row)
end.
```

This example takes the 2×3 two-dimensional array:

```
{{1, 2, 3},
 {4, 5, 6}}
```

and transposes it into the 3×2 array:

```
{{1, 4},
 {2, 5},
 {3, 6}}
```

This example demonstrates the use of arrays, including the index notation *Matrix*[*I*] and the new_array function. The first foreach loop uses two range iterators to transpose the *Matrix* array. If a foreach loop has multiple iterators, then it is compiled into a series of nested foreach loops in which each nested loop has a single iterator. In other words, a foreach loop with multiple iterators executes its goal once for every possible combination of values in the iterators. The first foreach loop corresponds to

```
foreach (I in 1..N)
   foreach (J in 1..M)
      Transposed[J,I] = Matrix[I,J]
   end
end
```

The second foreach loop iterates through the transposed array, printing one row at a time.

A loop statement forms a name *scope*. Variables that occur only in a loop, but do not occur before the loop in the outer scope, are local to each iteration of the loop. For example, in the following rule:

```
main =>
   A = {1,2,3},
   foreach (I in 1 .. length(A))
      E = A[I],
      writeln(E)
   end.
```

the variables I and E are local, and each iteration of the loop has its own values for these variables. Note that, like a loop statement, a list comprehension also forms a name scope.

Loops are compiled into tail-recursive predicates. Section 1.4.1 describes tail-recursive predicates.

1.4 Predicates and Functions

In Picat, predicates and functions are defined with *pattern-matching rules*. Each rule is terminated by a dot (.) followed by a white space.

1.4.1 Predicates

A *predicate* defines a relation, and can have zero, one, or multiple answers. Picat has two types of pattern-matching rules to define predicates: the *non-backtrackable* rule:

> *Head* => *Body*

or

> *Head* , *Cond* => *Body*

and the *backtrackable* rule:

> *Head* ?=> *Body*

or

> *Head* , *Cond* ?=> *Body*

Within a predicate, the *Head* is a *pattern* in the form $p(t_1, \ldots, t_n)$, where p is called the predicate *name*, each t_i is a *parameter* or an *argument* and n is called the *arity*. When $n = 0$, the parentheses can be omitted. The condition *Cond*, which is an optional goal, specifies a condition under which the rule is applicable. *Cond* cannot succeed more than once. A predicate is said to be *deterministic* if it is defined with only non-backtrackable rules, and *non-deterministic* if at least one of its rules is backtrackable. Note that, as shorthand, predicates and functions can be described as name/arity. For example, the length function has one parameter, and can be described as length/1, while the put function, discussed in Sect. 1.2.3, has three parameters, and can be described as put/3.

For a call C, if C matches the pattern $p(t_1, \ldots, t_n)$ and *Cond* is true, then the rule is said to be *applicable* to C. When applying a rule to call C, Picat rewrites C into *Body*. If the used rule is non-backtrackable, then the rewriting is a commitment, and the program can never backtrack to C. However, if the used rule is backtrackable, then the program will backtrack to C once *Body* fails, meaning that *Body* will be rewritten back to C, and the next applicable rule will be tried on C. In order to backtrack within the Picat interpreter, once a result is returned, the user can type a semicolon (;) to retrieve the next possible result, if there is such a result.

A pattern can contain *as-patterns* in the form *V@Pattern*, where *V* is a new variable in the rule, and *Pattern* is a non-variable term. The as-pattern *V@Pattern* is the same as *Pattern* in pattern matching, but after pattern matching succeeds, *V* is made to reference the term that matched *Pattern*. As-patterns can avoid reconstructing existing terms. For example,

```
p(V@[a|X]) => q(V).
```

is the same as

```
p([a|X]) => q([a|X]).
```

The difference is that in the former rule, the term `[a|X]` is not recreated, but in the latter rule, the term `[a|X]` is recreated.

Example: Min and Max of List

```
min_max1(List, Min, Max), List = [First|Rest] =>
    Mn = First,
    Mx = First,
    foreach (Value in Rest)
       if Value < Mn then
          Mn := Value
       elseif Value > Mx then
          Mx := Value
       end
    end,
    Min = Mn,
    Max = Mx.
```

In this example, the `min_max1` predicate returns the minimum value of `List` in `Min`, and returns the maximum value of `List` in `Max`. The condition `List = [First|Rest]` ensures that the predicate is not passed an empty list, and separates the head of the list from the tail. This allows the head to be stored as the initial minimum and maximum values. Then, the foreach loop cycles through each value in the list's tail, using an if-then-elseif statement to determine whether the value is less than the current minimum or greater than the current maximum. If either of these tests succeeds, the assignment operator `:=` updates the corresponding variable.

The following is a sample call to the `min_max1` predicate:

```
min_max1([17, 3, 41, 25, 8, 1, 6, 40], Min, Max)
```

The following predicate can also be used to find the minimum and maximum values of a list:

```
min_max2(List, Min, Max), List !== [] =>
    Min = min(List),
    Max = max(List).
```

Instead of using a foreach loop and an if-then-elseif statement, the `min_max2` predicate uses two of Picat's built-ins. The built-in min(*List*) finds the smallest value in a list. The built-in max(*List*) finds the largest value in a list. Note that the condition `List !== []` ensures that the `min_max2` predicate is not passed an empty list.

Example: The `zip` Function

The function zip(*List*₁, *List*₂, ..., *List*ₙ) returns a *zipped list* of array tuples from the argument lists, where *n* can be 2, 3, or 4. The `zip/2` function is implemented as follows:

```
zip([],_) = [].
zip(_,[]) = [].
zip([X|Xs],[Y|Ys]) = [{X,Y}|zip(Xs,Ys)].
```

The `zip` function can be used in an iterator in order to iterate over two or more lists simultaneously. For example, consider the following code:

```
main =>
    printf("The zipped list: %w%n", zip(1..2,3..4)),
    foreach ({X,Y} in zip(1..2,3..4))
        printf("%d+%d = %d%n",X,Y,X+Y)
    end.
```

This has the output:

```
The zipped list: [{1,3},{2,4}]
1+3 = 4
2+4 = 6
```

Within the foreach loop, the Picat compiler generates code for iterating over the list of pairs without actually computing the zipped list.

Example: Implementation of Member/2

A rule is said to be *tail-recursive* if the last call of the body is the same predicate as the head. The *last-call optimization* enables last calls to reuse the stack frame of the head predicate if the frame is not protected by any choice points. This optimization is especially effective for tail-recursion, because it converts recursion into iteration. Tail-recursion runs faster and consumes less memory than non-tail-recursion. For example:

```
member(X,[Y|_]) ?=> X=Y.
member(X,[_|L]) => member(X,L).
```

This example demonstrates the implementation of Picat's built-in member predicate. member(*Term*, *List*) determines whether *Term* is an element of *List*. The first rule of the `member` predicate is backtrackable, allowing *Term* to be instantiated to different elements of the list. The second rule is tail-recursive, because the end of the rule calls the `member` predicate again.

1.4.2 Predicate Facts

For an extensional relation that contains a large number of Prolog-style facts, it is tedious to define such a relation as a predicate with pattern-matching rules. In order to facilitate the definition of extensional relations, Picat allows the inclusion of *predicate facts* in the form $p(t_1, \ldots, t_n)$ in predicate definitions. Facts and rules cannot co-exist in predicate definitions, and facts must be ground. A predicate definition that consists of facts must be preceded by an *index declaration* in the form

$$\text{index } (M_{11}, M_{12}, \ldots, M_{1n}) \ldots (M_{m1}, M_{m2}, \ldots, M_{mn})$$

where each M_{ij} is either $+$ (meaning indexed) or $-$ (meaning not indexed). Facts are translated into pattern-matching rules before they are compiled.

Example: Predicate Facts

```
index (+,-) (-,+)
edge(a,b).
edge(a,c).
edge(b,c).
edge(c,b).
```

The predicate `edge`, which represents edges in a graph, is translated into the following rules:

```
edge(a,Y)  ?=>  Y=b.
edge(a,Y)  =>   Y=c.
edge(b,Y)  =>   Y=c.
edge(c,Y)  =>   Y=b.
edge(X,b)  ?=>  X=a.
edge(X,c)  ?=>  X=a.
edge(X,c)  =>   X=b.
edge(X,b)  =>   X=c.
```

The `edge` predicate can be queried as follows:

```
Picat> edge(a,X), edge(X,Y)
X = b
Y = c;
X = c
Y = b
```

Note the semicolon after `Y = c`. When the user types the semicolon, the system backtracks, finding a second possible solution. This is an example of a multi-value query, which finds two results through backtracking: `X = b, Y = c` and `X = c, Y = b`.

1.4.3 Functions and Function Facts

A *function* is a special kind of a predicate that always succeeds with *one* answer. In order to define functions, Picat uses the pattern-matching rule:

$$Head = Exp => Body$$

or

$$Head = Exp, \ Cond => Body$$

Within a function, the *Head* is an equation $f(t_1, \ldots, t_n) = X$, where f is called the *function name*, and X is an *expression* that gives the return value. Functions are defined with non-backtrackable rules only.

For a call C, if C matches the pattern $f(t_1, \ldots, t_n)$ and *Cond* is true, then the rule is said to be *applicable* to C. When applying a rule to call C in the equation $Res = C$, Picat rewrites the equation into $(Body, \ Res = Exp)$.

Picat allows inclusion of *function facts* in the form $f(t_1, \ldots, t_n) = Exp$ in function definitions. The function fact $f(t_1, \ldots, t_n) = Exp$ is shorthand for the rule

$$f(t_1, \ldots, t_n) = X => X = Exp$$

where X is a new variable.

Although all functions can be defined as predicates, it is preferable to define them as functions for two reasons. Firstly, functions often lead to more compact expressions than predicates, because arguments of function calls can be other function calls. Secondly, functions are easier to debug than predicates, because functions never fail and never return more than one answer.

Example: Quicksort

```
qsort([]) = [].
qsort([H|T]) = L =>
    L = qsort([E : E in T, E=<H]) ++
        [H] ++
        qsort([E : E in T, E>H]).
```

This example performs the *quicksort* algorithm to sort a list. The first rule is a function fact. The second rule is a function that concatenates three lists. Note that the second rule contains two recursive calls.

The following is a sample call to the `qsort` function:

```
L = qsort([17, 3, 41, 25, 8, 1, 6, 40])
```

Note that a function call cannot be used as a goal, even if the function returns `true` or `false`. Instead, a function call must be used in an expression.

1.4.4 Introduction to Tabling

A predicate defines a relation where the set of facts is implicitly generated by the rules. The process of generating the facts may never end and/or may contain a lot of redundancy. Tabling can prevent infinite loops and redundancy by memorizing calls and their answers. In order to have all calls and answers of a predicate or function tabled, users just need to add the keyword `table` before the first rule.

Example: Fibonacci Numbers

```
table
fib(1) = 1.
fib(2) = 1.
fib(N) = F, N > 2 => F = fib(N-1)+fib(N-2).
```

This example demonstrates a function $fib(N)$ that finds the Nth *Fibonacci number*. Fibonacci numbers are recursively defined as: `f(1) = 1`, `f(2) = 1`, `f(N) = f(N-1) + f(N-2)`. When not tabled, the function call $fib(N)$ takes exponential time in N. However, when tabled, it only takes linear time.

For more details on tabling, see Chap. 4.

1.5 Picat Compared to Other Languages

Picat integrates several programming paradigms, including logic programming, imperative programming, functional programming, scripting, dynamic programming with tabling, and constraint programming. This section compares Picat with Prolog as a logic programming language, Haskell as a functional programming language, and Python as a scripting language.

1.5.1 Picat Versus Prolog

Although Picat is a multi-paradigm language, its core is underpinned by logic programming concepts, including *logic variables*, *unification*, and *backtracking*.

Like in Prolog, logic variables in Picat are value holders. A logic variable can be bound to any term, including another logic variable. Logic variables are single-assignment, meaning that once a variable is bound to a value, the variable takes the identity of the value, and the variable cannot be bound again, unless the value is a variable or contains variables.

In both Prolog and Picat, unification is a basic operation, which can be utilized to unify terms. Unlike Prolog, Picat uses pattern-matching, rather than unification, to select applicable rules for a call. Consider the predicate `membchk/2`, which checks whether a term occurs in a list. The following are its definitions in Prolog and in Picat:

```
% Prolog
membchk(X,[X|_]) :- !.
membchk(X,[_|T]) :- membchk(X,T).
```

```
% Picat
membchk(X,[X|_]) => true.
membchk(X,[_|T]) => membchk(X,T).
```

For a call `membchk(X,L)`, if both `X` and `L` are ground, then the call has the same behavior under both the Prolog and the Picat definitions. However, if `X` or `L` is not ground, then the call may give different results in Prolog and in Picat. For example, the call `membchk(a,_)` and the call `membchk(_,[a])` succeed in Prolog, but they fail in Picat. Pattern-matching is a simpler operation than unification. Pattern-matching rules in Picat are fully indexed, while Prolog clauses are typically indexed on one argument. Therefore, Picat can be more scalable than Prolog.

Picat, like Prolog, supports backtracking. In Prolog, every clause is implicitly backtrackable, and the *cut operator* `!` can be utilized to control backtracking. In contrast, backtrackable rules in Picat must be explicitly denoted, which renders the cut operator unnecessary. Consider the predicate `member/2`, as defined in Prolog and in Picat:

```
% Prolog
member(X,[X|_]).
member(X,[_|T]) :- member(X,T).
```

```
% Picat
member(X,[Y|_]) ?=> X = Y.
member(X,[_|T]) => member(X,T).
```

For a call `member(X,L)`, if `L` is a *complete list*,[4] then the call has the same behavior under both the Prolog and the Picat definitions. If `L` is incomplete, then the call can

---

[4]A list is *complete* if it is empty, or if its tail is complete. For example, `[a,b,c]` and `[X,Y,Z]` are complete, but `[a,b|T]` is not complete if `T` is a variable.

succeed an infinite number of times under the Prolog definition, because unifying L with a cons always succeeds when L is a variable. In contrast, pattern matching never changes call arguments in Picat. Therefore, the call member(X,L) can never succeed more times than the number of elements in L.

Picat also differs from Prolog in several other aspects, such as the lack of support for dynamic predicates and for user-defined operators in Picat. The most distinguishable difference between Prolog and Picat is in Picat's language constructs, including list and array comprehensions, assignable variables, loops, and functions. These language constructs, together with the built-in array and map data types, make Picat more convenient than Prolog for scripting and modeling tasks. This book is full of examples for constraint solving and planning that benefit from these constructs.

The following gives an example function in Picat, which takes two matrices, A and B, and returns the product $A \times B$:

```
matrix_multi(A,B) = C =>
   C = new_array(A.length,B[1].length),
   foreach (I in 1..A.length, J in 1..B[1].length)
      C[I,J] = sum([A[I,K]*B[K,J]
                      : K in 1..A[1].length])
   end.
```

Prolog's definition of the same function would be 10 times as long as Picat's definition. Furthermore, since the Picat compiler translates loops and list comprehensions into tail recursion, which is further converted into iteration by tail-recursion optimization, the Picat version is as fast as the Prolog version, if not faster.

1.5.2 Picat Versus Haskell

Like Haskell, Picat supports function definitions with pattern-matching rules. A major difference between Haskell and Picat is that Haskell is a statically-typed language, while Picat is a dynamically-typed language. In Haskell, every variable has a known type at compile time, while in Picat, a variable is typeless until it is bound to a value. Although static typing allows the Haskell compiler to detect type errors and to optimize code generation by automatically inferring types, Picat is more flexible than Haskell.

Haskell is a pure functional language, while the core of Picat is a logic language. Haskell and Picat, as two *declarative programming languages*, discourage the use of side effects in describing computations. All of the built-in functions in Picat's basic module are side-effect-free mathematical functions. For example, the sort(*L*) function returns a sorted copy of list *L* without changing *L*, and the remove_dups(*L*) function returns a copy of *L* that has no duplicate values. Pure, side-effect-free functions are not dependent on the context in which they are applied. This purity can greatly enhance the readability and maintainability of programs.

In Haskell, impure computations that have side effects are described as *monads*. Haskell's type system distinguishes between pure and *monadic* functions. In Picat, side effects can be caused by the assignment operator : = and by certain built-ins, such as those that perform I/O. An assignment of the form S[I] := RHS has global side effects, since the compound term S is destructively updated, like an assignment in an imperative language. An assignment of the form X := RHS, where X is a variable, only has a side effect within the body of the rule in which the assignment occurs. Recall that the compiler introduces a new variable for X and replaces the remaining occurrences of *X* by the new variable. Variable assignments do not have cross-predicate or cross-function side effects.

In Haskell, using *higher-order functions* is a basic norm of programming. Although Picat has limited support for higher-order predicates, such as call/n, and functions, such as apply/n, map/2, and fold/3, the use of higher-order calls is discouraged, because of the overhead. Whenever possible, recursion, loops, or list and array comprehensions should be used instead.

In Picat, the unification operator = and the equality operator == can be used on any two terms. In contrast, in Haskell, the equality operator == can only be used on certain types of values. For example, consider the membchk function in Haskell:

```
membchk x (y:_)
     | x == y    = True
     | otherwise = False
membchk _ [] = False
```

This function has the type (Eq a) => a -> [a] -> Bool, which requires type a to be an instance of a typeclass that defines the equality operator.

Haskell supports implicit *lazy evaluation*, meaning that the evaluation of an expression can be delayed until the value of the expression is needed. In contrast, Picat is a strict language, meaning that arguments are completely evaluated before calls are evaluated. Lazy evaluation allows for the creation of infinite lists. For example, the Haskell expression [1,2..] creates an infinite list of positive integers. In Picat, the freeze(X,Call) predicate, which delays Call until X is instantiated, can be used to simulate lazy evaluation, and infinite data can be created through the use of backtracking. For example, the following predicate generates an infinite sequence of positive integers:

```
gen(X) ?=> X =1.
gen(X) => gen(X1), X = X1+1.
```

Picat is essentially a relational language. Logic variables and automatic backtracking are two features that make Picat more suitable for search problems than Haskell. The built-in predicate append is probably one of the most powerful and convenient nondeterministic predicates in Picat. The following examples illustrate several different uses of this predicate:

```
append(Pre,_,L)              Checks if Pre is a prefix of L
append(_,Suf,L)              Checks if Suf is a suffix of L
append(_,[Last],L)           Retrieves the last element of L
append(Part1,[Sep|Part2],L) Splits L into 2 parts separated by Sep
```

Logic variables are easily extended to domain variables, which enable the natural specification of constraints. Although it is possible to simulate backtracking by using lazy evaluation in Haskell, the native support of backtracking is an advantage of Picat for relational programming.

1.5.3 Picat Versus Python

Both Python and Picat are dynamically-typed and interpreted languages that emphasize flexibility and brevity of description more than the efficiency of programs.

Python is an imperative language and many of its built-ins have side effects. For example, the `lst.sort()` method destructively sorts the list `lst`, and the `lst.remove(elm)` method destructively removes `elm` from `lst`. Side effects can make code less manageable.

Lists in Picat, like lists in Prolog and Haskell, are linked lists, while lists in Python are dynamic arrays. In Picat, for a list L, the function call `len(L)` takes linear time in the size of L, and the list access `L[I]` takes *I* steps to reach the *I*th element. In contrast, in Python, both `len(L)` and `L[I]` take constant time. Picat supports constant-time access of array elements, but Picat's arrays are not dynamic.

Python does not support tail-recursion optimization. Because of this, iteration is more often favorable than recursion for describing repetitions. For example, consider the following definition of `membchk` in Python:

```
def membchk(elm, lst):
    if lst == []:
        return False
    else:
        [head,*tail] = lst
        if head == elm:
            return True
        else:
            return membchk(elm,tail)
```

This definition is not efficient since: (1) the assignment `[head,*tail] = lst` creates a new list object, and (2) the recursive call `membchk(elm,tail)` creates a new frame on the stack, rather than reusing the caller's stack frame. A more efficient implementation is to use a loop to iterate over the list.

As can be seen in the `membchk` function shown above, Python allows patterns on the left-hand sides of assignments. While pattern-matching is a nice feature of Python, pattern-matching rules in Picat (and in Haskell) are more easily manageable and efficient. Each pattern-matching rule has its own naming scope, which enables the addition and removal of rules without affecting other rules. Pattern-matching rules are fully indexed, and in many cases pattern-directed branching is much more efficient than if-then-else.

Unlike Python, Picat does not support object-orientation. It is possible to simulate abstract data types in Picat using the module system. For example, a rectangle class can be defined in a module that provides a function named `new_rect` for creating a rectangle structure, functions for getting and setting the attribute values of a rectangle, and some other functions, including a function that computes a rectangle's area. Picat's dot notation makes calling a function on a structure look like calling a method on an object. For example, the following query creates a rectangle and computes its area:

```
R = new_rect(100,100), Area = R.area().
```

More syntax extensions are necessary to make Picat a full-fledged OOP language.

As a logic language, Picat has the same advantages over Python as it has over Haskell for search problems. Python has been used as a host language for constraint programming and linear programming solvers. Logic variables tend to make Picat models neater than their Python counterparts. Backtracking makes it convenient to implement some search strategies, such as branch-and-bound and iterative-deepening, for the underlying solvers.

1.6 Writing Efficient Programs in Picat

This section gives several Picat-specific tips for writing efficient programs.

1.6.1 Making Definitions and Calls Deterministic

Backtracking is a nice feature of Picat, which facilitates describing and solving search problems. However, backtracking is not cost-free. A call to a nondeterministic predicate uses a frame that stores not only the call's arguments and the caller's environment, but also the information needed for backtracking. As long as the call has rules that have not yet been applied, the frame stays on the stack. For this reason, it is always good practice to make definitions deterministic, if possible, by using the non-backtracking rule *LHS => RHS* instead of the backtrackable rule *LHS ?=> RHS*.

The `once` operator can be utilized to inform the system that a call is required to succeed at most once. For example, consider the call

```
append(Str1,[Sep|Str2],Str).
```

One of the uses of this call is to split the string `Str` into two substrings, `Str1` and `Str2`, that are separated by `Sep`. For example,

```
Picat> append(Str1,[' '|Str2],"hello world Picat")
Str1 = [h,e,l,l,o]
Str2 = [w,o,r,l,d,' ','P',i,c,a,t] ?;

Str1 = [h,e,l,l,o,' ',w,o,r,l,d]
Str2 = ['P',i,c,a,t] ?
```

If only one answer is needed, then the call should be written as

```
once append(Str1,[Sep|Str2],Str).
```

After the call succeeds, the system deallocates the call's frame. The use of `once` can lead to better efficiency, because it prevents the system from performing unnecessary backtracking.

Some of Picat's nondeterministic built-ins have deterministic counterparts. For example, while `member(X,L)` can succeed multiple times, `membchk(X,L)` can succeed at most once. If the purpose of the call is to check if a value `X` is a member of a list, then `membchk(X,L)` should be used, because it is generally more efficient than `once member(X,L)`.

1.6.2 Making Definitions Tail-Recursive

The Picat compiler adopts an optimization technique, called *last-call optimization*, which enables the last call in a rule's body to reuse the caller's stack frame under certain circumstances. This optimization is especially effective for tail-recursive definitions, since it essentially converts recursion into iteration. Consider the following function:

```
length([]) = 0.
length([_|T]) = length(T)+1.
```

The second rule is recursive, but not tail-recursive, because the last function call in the body is addition. One technique for converting non-tail-recursive definitions into tail-recursive ones is to use an extra argument which *accumulates* a value during recursion and returns the value when the base case is reached. For example, the `length/2` predicate can be converted into the following tail-recursive definition:

```
length(L) = length(L,0).

length([],Len) = Len.
length([_|T],Len) = length(T,Len+1).
```

The second argument of `length/2` accumulates the number of elements that have been scanned so far. In the beginning, the accumulator has the value 0. After each element is scanned, the accumulator is incremented. When the list becomes empty, the accumulator is returned as the length of the original list.

1.6.3 Incremental Instantiation and Difference Lists

Incremental instantiation means passing a non-ground term to a predicate and letting the predicate instantiate some of the variables in the term. The resulting term can be passed to another predicate. In this way, computation can be done by incrementally instantiating terms.

Incremental instantiation can be used to construct lists. For example, the following sequence of unification calls incrementally creates the list `[a,b,c]`:

```
L = [a|L1], L1 = [b|L2], L2 = [c|L3], L3 = []
```

The statement `L = [a|L1]` can be viewed as calling a predicate that takes `L` and `L1` as parameters and creates the list `[a]` as a *difference list*, `L-L1`. Similarly, the call `L1 = [b|L2]` creates the list `[b]` as the difference `L1-L2`. This sequence of unification calls is as efficient as the call `L = [a,b,c]`, and is more efficient than `[a] ++ [b] ++ [c]`.

Difference lists are useful for writing tail-recursive predicates, and can sometimes reduce the time and space complexity by eliminating the need to call `++`. For example, consider the following function, which returns the flattened list of a nested list:

```
flatten([]) = [].
flatten([H|T]) = flatten(H) ++ flatten(T).
flatten(Any) = [Any].
```

The following gives an example use of this predicate:

```
Picat> FL = flatten([[[a],b],c]).
FL = [a,b,c]
```

Given the *nested list*, `[[[a],b],c]`, the predicate *flattens* it into the list `[a,b,c]`.

The above definition of `flatten` is not efficient, because: (1) it is not tail-recursive, and (2) it may take more than linear time to flatten a list. The following gives a more efficient definition:

```
flatten(L) = FL => flatten(L,FL,[]).

flatten([],FL,FLr) => FL = FLr.
flatten([H|T],FL,FLr) =>
    flatten(H,FL,FL1),
    flatten(T,FL1,FLr).
flatten(Any,FL,FLr) => FL = [Any|FLr].
```

The predicate call `flatten(L,FL,FLr)` returns the flattened list of `L` as the difference list `FL-FLr`. If `L` is empty, then the flattened list is also empty, meaning that `FL = FLr`. If `L` is the cons `[H|T]`, then the flattened list of `L` is `FL-FLr`, which consists of the flattened part of `H` (`FL-FL1`) and the flattened part of `T` (`FL1-FLr`).

1.6.4 Writing Efficient Iterators

In Picat, there are normally several different ways of describing the same repeated computation. Programmers should know which is the most efficient way of describing the computation. Consider the following two loop statements:

```
foreach (E in L)
    print(E)
end.

foreach (I in 1..length(L))
    print(L[I])
end.
```

If `L` is an array, then the two loops have the same performance. For an array `L`, `length(L)` and `L[I]` both take constant time to compute. If `L` is a list, then the first loop takes linear time, while the second loop takes quadratic time, because, when `L` is a list, both `length(L)` and `L[I]` take linear time to compute.

The following example accesses each element of a list by using its index:

```
foreach (I in 1..length(L))
    printf("L[%w] = %w\n",I,L[I])
end.
```

As explained above, this loop takes quadratic time. The following loop is more efficient:

```
foreach ({I,E} in zip(1..length(L),L))
    printf("L[%w] = %w\n",I,E)
end.
```

The Picat compiler generates code for iterating over the list of pairs without actually computing the zipped list. This loop still requires the computation of `length(L)` before it starts, but the overall running time is linear. If performance is critical, then the computation should be defined by using recursion, as follows:

```
loop_pred([],_I) => true.
loop_pred([E|L],I) =>
    printf("L[%w] = %w\n",I,E),
    loop_pred(L,I+1).
```

The original loop is equivalent to the predicate call `loop_pred(L,1)`.

1.7 Bibliographical Note

The details of the Picat language can be found in [72]. Picat shares many features with Prolog, especially B-Prolog [68]. While knowing Prolog is not a prerequisite to learning programming in Picat, knowledge about Prolog and its programming techniques is certainly helpful. Many textbooks on Prolog are available (e.g., [8, 14, 33, 53]). Most of the problem-solving and programming techniques described in these books can be easily adapted to Picat. The book by O'Keefe [42] describes several advanced programming techniques for Prolog, including exploitation of determinism, tail-recursion, and difference lists.

Several logic programming systems support language extensions, such as functions in Mercury [40] and Ciao [24], and loop constructs in ECLiPSe [49] and B-Prolog [68]. ECLiPSe requires the declaration of global variables, and B-Prolog requires the declaration of local variables. In contrast, Picat adopts a simple and clean scoping rule for variables, which renders the declaration of global and local variables unnecessary.

Assignable variables are rarely seen in logic and functional languages. The Picat compiler introduces a new variable for each assigned variable. This transformation is employed in building the static single assignment form (SSA) for imperative programs, which simplifies program analysis and compilation [16].

Some of Picat's constructs and notations are borrowed from other languages. The as-pattern is taken from Haskell [26]. The list comprehension, which can be traced back to SETL [50], was made popular by Haskell. Picat's dot-notation for chaining function calls is borrowed from object-oriented languages. Several textbooks on programming languages give a comprehensive comparison of different programming language paradigms (e.g., [51, 52]).

The Picat implementation adopts the virtual machine ATOAM [66], which is a redesign of the Warren Abstract Machine [63] for fast software emulation. The extended virtual machine that supports tabling and constraint propagation is detailed in [68].

1.8 Exercises

1. Write a program that prints a staircase of a given number of steps. For example, here is the output when the number is 3:

```
        +---+
        |   |
    +---+---+
    |   |   |
+---+---+---+
|   |   |   |
+---+---+---+
```

2. Write a predicate `pascals_triangle(N)` that prints the first N rows of *Pascal's triangle*. Starting with row 0, each row, I, has $I + 1$ columns, defined by `Triangle[I, J] = I! / (J! * (I - J)!)`, where ! is shorthand for *factorial*. For example, the following shows rows 0 through 3 of Pascal's triangle:

```
    1
   1 1
  1 2 1
 1 3 3 1
```

   Hint: remember that lists and arrays begin with index 1.

3. The transpose example in Sect. 1.3.4 is defined in the `main` predicate, and transposes a predetermined 2×3 array into a 3×2 array. Convert this example into a function `transpose(Matrix)` = *Transpose* that takes an arbitrary 2-dimensional array and returns the array's transpose instead of printing it within the function.

4. Write a function `number_of_zeroes/1` that returns the number of zeroes that are in a given list.

5. Write a function `all_permutations/1` that returns a list of permutations of a given list. For example, `all_permutations([1,2,3])` returns:

```
[[1,2,3],[1,3,2],[2,1,3],[2,3,1],[3,1,2],[3,2,1]]
```

6. Modify the quicksort example from Sect. 1.4.3 so that it uses two function facts, instead of one function fact and one function.

7. Convert the Fibonacci example from Sect. 1.4.4 to only use a single rule, which contains an in-then-else statement.

8. Convert the Fibonacci example from Sect. 1.4.4 to find the Nth *Tribonacci number*. Tribonacci numbers are defined as: `t(1) = 1, t(2) = 1, t(3) = 2, t(N) = t(N-1) + t(N-2) + t(N-3)`.

9. The following call prints a list of prime numbers from 2 to 100:

```
print([P : P in 2..100,
       foreach (M in 2..P-1) P mod M > 0 end])
```

Write a program to do the same without using any loops or list comprehensions.
10. Convert the following into tail-recursive functions:

(a) Reverse a list:

```
rev([]) = [].
rev([H|T]) = rev(T)++[H].
```

(b) Sum a list:

```
sm([]) = 0.
sm([H|T]) = sm(T)+H.
```

Chapter 2
Basic Constraint Modeling

Abstract Given a set of variables, each of which has a domain of possible values, and a set of constraints that limit the acceptable set of assignments of values to variables, the goal of a CSP (Constraint Satisfaction Problem) is to find an assignment of values to the variables that satisfies all of the constraints. Picat provides three solver modules, including `cp` (Constraint Programming), `sat` (Satisfiability), and `mip` (Mixed Integer Programming), for modeling and solving CSPs. This chapter provides an introduction to modeling with constraints, with a primary focus on the `cp` module, and a secondary focus on the `sat` and `mip` modules.

2.1 Introduction

In Picat, it is possible to use the same model and syntax for three different solver modules. The `cp` and `sat` modules support integer-domain variables, while the `mip` module supports both integer-domain and real-domain variables. This chapter introduces constraint modeling for solving CSPs (Constraint Satisfaction Problems), in which the possible values of the domains are integers which are often not too large.

Note: The models that use the `sat` solver do not work in the Windows version of Picat, because the Lingeling SAT solver doesn't compile with MVC. Windows users can install Cygwin,[1] and use the Cygwin version instead.

In general, a constraint model consists of a set of *decision variables*, each of which has a specified domain, and a set of *constraints*, each of which restricts the possible combinations of values of the involved decision variables. In order to use a solver, a constraint program must first import the solver module. A constraint program normally poses a problem in three steps: (1) generate variables; (2) generate constraints over the variables; and (3) call `solve` to invoke the solver in order to find a valuation for the variables that satisfies the constraints and possibly optimizes an objective function.

A CSP forms a search space. Solvers have techniques for reducing the search space, and use strategies to search the space for solutions. For example, CP and SAT solvers use *constraint propagation* to prune unfruitful paths in the search space.

---

[1] https://www.cygwin.com/.

© The Author(s) 2015
N.-F. Zhou et al., *Constraint Solving and Planning with Picat*,
SpringerBriefs in Intelligent Systems, DOI 10.1007/978-3-319-25883-6_2

A decision variable is also called a *domain variable*. In Picat, the domain constraint X :: D can be used to restrict the domains of variables. A domain variable is a special kind of a logic variable. General logic variables are assumed to have the set of all possible ground terms, called the *Herbrand base*, as their implicit domains, and operations such as unification can be used to restrict their possible values. In contrast, domain variables have explicit domains, and a rich set of built-in constraints, such as arithmetic constraints and the `all_different` constraint, can be used to restrict their possible values.

The three solver modules in Picat have different strengths and weaknesses. The `cp` module tends to be the best choice for problems in which effective global constraints and/or problem-specific labeling strategies are available. The `sat` module is often well-suited to problems that can be clearly represented as Boolean expressions or have efficient CNF (Conjunctive Normal Form) encodings. The `mip` module is still the best choice for many kinds of Operations Research problems. It is often instructive to test all three solvers on the same problem. The common interface that Picat provides for the solver modules allows seamless switching from one solver to another.

The programs in this chapter and in the next chapter demonstrate an important aspect of constraint modeling, namely *declarative programming*. The ideal is to just state the problem and let the solver do the job, at least in principle. Logic programming also has this feature, but, arguably, constraint modeling takes this one step (or several steps) further. Depending on the specific problem at hand, it might still be necessary to explicitly state certain things, such as for loops and list comprehensions, so please take the notion of "declarativity" with a grain of salt.

2.2 SEND+MORE=MONEY

SEND+MORE=MONEY is one of the most commonly-used examples when introducing CSPs. The problem is to substitute each letter (SENDMORY) with a distinct digit in the range of 0..9, such that the equation SEND + MORE = MONEY is satisfied. The following gives a model for the problem. Another model, which uses carrying in the same way as multiplication by hand, is left as an exercise.

```
import cp.

main =>
   Digits = [S,E,N,D,M,O,R,Y],
   Digits :: 0..9,
   all_different(Digits),
   S #> 0,
   M #> 0,
                  1000*S + 100*E + 10*N + D
   +              1000*M + 100*O + 10*R + E
```

```
#= 10000*M + 1000*O + 100*N + 10*E + Y,
solve(Digits),
println(Digits).
```

Here is a breakdown of the code:

`import cp`: This ensures that the CP solver is used. The module name `cp` can be changed to `sat` or `mip` to use a different solver.

`Digits :: 0..9`: This defines the domains of the decision variables in the list `Digits`, and thus the single variables S, E, N, D, M, O, R, and Y. Recall that the range `0..9` denotes the list of integers from 0 through 9.

`#>` and `#=`: These are *arithmetic constraint operators.* The disequality constraints S `#>` 0 and M `#>` 0 ensure that the leading digits of SEND and MORE must both be greater than 0. The equality constraint ensures that SEND plus MORE equals MONEY. Note that all of the arithmetic constraint operators begin with #. This special notation distinguishes between the constraint operators and the normal relational operators in Picat.

Functions in arithmetic constraints are treated as data constructors, with the exception of list comprehensions and index notations. In this example, the expressions on the two sides of the equality constraint are constructed as two terms before being passed to the predicate that defines `#=`. Changing `#=` in the program to = would result in an evaluation error, since the expressions involve uninstantiated variables, meaning that the functions cannot be evaluated.

`all_different(Digits)`: `all_different` is a *global constraint* which states that all of the decision variables in `Digits` must be distinct. In contrast to unary or binary arithmetic constraints that have one or two arguments, global constraints are often defined to work on a larger collection of decision variables, such as a list. Global constraints are quite unique to CP, compared to other types of CSP solvers, since the CP solver employs special propagation algorithms for global constraints in order to reduce the search space. For this reason, global constraints are often high-level *tools-for-thought* for modeling problems with CP.

Global constraints can be decomposed into smaller constraints. For example, `all_different` (*L*) can be decomposed into a set of inequality constraints that contains *X* `#!=` *Y* for each pair of distinct elements *X* and *Y* in *L*. However, these decompositions are often slower than the specially-crafted global constraints.

`solve(Digits)`: The `solve` predicate finds an assignment of values to the variables that satisfies all of the accumulated constraints. In general, all of the decision variables should be passed to `solve`.

2.3 Sudoku—Constraint Propagation

This section shows how to model the *Sudoku problem*, using a small 4 × 4 Sudoku instance to demonstrate how constraint propagation changes the domains of the variables during search. Note that the behavior that is described in this section is specific to the CP solver. The SAT and MIP solvers do not use domain reduction in the same manner.

2.3.1 A Sudoku Program

Given an $N \times N$ board with some squares already filled with hint values, the objective of the Sudoku problem is to fill all the empty squares with values in $1..N$, such that each row, each column, and each of the N blocks are filled with distinct values. Figure 2.1 gives a Picat program for the problem.

The predicate board(Board) gives a problem instance. A board configuration is given as a matrix (i.e., a 2-dimensional array), where an anonymous variable, indicated by _ (an underscore), means an unknown, and the numbers are pre-filled hint values. The predicate sudoku/1 first calculates the dimension N and the block size N1 of the instance. For the given instance, $N = 4$ and $N1 = 2$. The domain constraint Board :: 1..N changes all the anonymous variables to domain variables with the domain 1..N. The predicate sudoku/1 then generates an all_different constraint for each row, each column, and each of the N blocks. Finally, it calls the solver to *label* the variables with values, meaning that the solver assigns values to the variables.

Modeling tip: The row and column constraints and the list comprehensions should now be familiar. The third constraint, the list comprehension for the block constraints, might take some time to figure out, both to model and to understand. One tip when modeling this kind of constraint is to first do a "pen and paper version" for finding the proper indices and to then model the indices that are needed for each block.

Note that this problem instance has two solutions. Usually, a Sudoku problem has a single, unique solution.

2.3.2 Constraint Propagation—Concepts

One of the key operations employed in a CP solver is called *constraint propagation*, which actively uses constraints to prune search spaces. Whenever a variable changes, meaning that the variable has been instantiated or a value has been excluded from its domain, the domains of all the remaining variables are filtered to contain only those values that are consistent with the modified variable. Section 2.3.3 provides a concrete example of this domain reduction. There may exist different propagation rules for a constraint, depending on the level of *consistency* to be achieved, i.e., on the efforts to be made to reduce domains.

A *fixpoint* is reached after the propagation phase when no domains can be reduced any further. A fixpoint represents a *failure* if any of the domains becomes empty, and a fixpoint represents a solution if all of the domains become *singletons*, i.e.,

```
import cp.

main =>
  board(Board),
  sudoku(Board),
  println(Board),
  nl.

sudoku(Board) =>
  N = Board.length,        % dimension of Board
  N1 = ceiling(sqrt(N)),   % block size
  Board :: 1..N,

  % row constraints
  foreach (I in 1..N)
    all_different([Board[I,J] : J in 1..N])
  end,

  % column constraints
  foreach (J in 1..N)
    all_different([Board[I,J] : I in 1..N])
  end,

  % block constraints
  foreach (I in 1..N1..N, J in 1..N1..N)
    all_different([Board[I+K,J+L] :
                   K in 0..N1-1, L in 0..N1-1])
  end,
  solve(Board).

board(Board) =>
  Board = {{4, _,  _, _},
           {_, 1,  _, _},
           {_, _,  _, 1},
           {_, _,  _, 2}}.
```

Fig. 2.1 A program for Sudoku

single-element domains. In general, the propagation phase may not be able to reach a failure or a solution, in which case the CP solver needs to use another key operation, called *search* or *labeling*, to guess a value or a range of values for a variable. After a guess, constraint propagation reduces the domains of the related variables. When a failure is reached, the CP solver *backtracks* to a previous guess and tries a different value or values for the variable.

The performance of a constraint program is greatly affected by *search strategies*, which decide the variables to select and the values to guess. In Picat, search strategies can be given to the CP solver as options in solve (*Options*, *Vars*). The ff (*first-fail*) strategy, which selects the leftmost variable with the smallest domain, is often used.

2.3.3 Constraint Propagation—Example

Before reading further, try to solve this Sudoku instance by hand, and study the thinking used to reach the different values. For instance, one may think, "since this is the only possible value in this cell, the cell must be that value." One might also think, "the values in these rows and columns, when taken together, mean that this cell must have that value." Reasoning like this is a way of *reducing the domains* of the different variables, as mentioned in Sect. 2.3.2.

For example, the following shows how constraint propagation changes the domains of the variables after each loop statement in the Sudoku program. Initially, after `Board::1..N`, all of the domains are either a hint value or the full domain (`1..4`).

```
     4          1..4          1..4          1..4
   1..4           1           1..4          1..4
   1..4          1..4          1..4           1
   1..4          1..4          1..4           2
```

After the loop that posts the *row constraints*, the domains change to the following:

```
     4          1..3          1..3          1..3
   2..4           1           2..4          2..4
   2..4          2..4          2..4           1
  1,3,4         1,3,4         1,3,4           2
```

Since the hint value at cell [1,1] is 4, the `all_different` constraint for the first row excludes 4 from the domains at cells [1,2], [1,3], and [1,4]. The domains in the other rows are changed in a similar fashion.

After the loop that posts the *column constraints*, the domains change to the following:

```
     4            2             1             3
    2,3           1            2,3            4
    2,3          3,4           2..4           1
    1,3          3,4           3,4            2
```

The hint value at cell [3,4] is 1, and the value at cell [4,4] is 2. After the column constraint for column 4 is posted, the domain at cell [1,4] changes to 3, and the domain at cell [2,4] changes to 4. After the domain at cell [1,4] changes to 3, the row constraint for the first row is activated to exclude 3 from the domains at cells [1,2] and [1,3]. These changes will trigger the other constraints to reduce the related domains before a fixpoint is reached.

After the loop that posts the *block constraints*, the domains change to the following:

4	2	1	3
3	1	2	4
2	3,4	3,4	1
1	3,4	3,4	2

The constraint for the upper-left block removes 2 from the domain at cell [2,1], the constraint for the upper-right block removes 3 from the domain at cell [2,3], and the constraint for the lower-right block removes 1 from the domain at cell [3,3]. These changes will trigger the row and column constraints to further reduce the related domains before a fixpoint is reached.

When `solve(Board)` is executed, there are only four uninstantiated variables. After guessing 3 for the domain at cell [3,2], the constraints are triggered to remove 3 from the domains at cell [3,3] and cell [4,2]. After these changes, 4 is removed from the domain at cell [4,3]. For this instance, a single guess is sufficient to find a solution, meaning that backtracking is not needed. Note that this behavior is more of an exception; most interesting problems normally require backtracking.

2.4 Minesweeper—Using SAT

The *Minesweeper problem* in this context is not the full GUI-based program that is normally associated with the name "Minesweeper". Instead, it is a simplified problem, where the goal is to identify the positions of all the mines in a given matrix, with hints that state how many mines there are in the neighboring cells, including diagonal neighbors. A cell in the middle of the matrix has eight neighbors, a cell in the corner has three neighbors, and an edge cell has five neighbors.

Here is a problem instance:

```
.  .  2  .  3  .
2  .  .  .  .  .
.  .  2  4  .  3
1  .  3  4  .  .
.  .  .  .  .  3
.  3  .  3  .  .
```

For this instance, the third cell in the first row has the value 2, which indicates that it has two adjacent mines. The cells marked with "." are unknowns, meaning that the cell may or may not have a mine. One can also observe that if there is a number in a cell, then the cell cannot contain a mine.

```
import sat.

main =>
  % define the problem instance
  problem(Matrix),
  NRows = Matrix.length,
  NCols = Matrix[1].length,

  % decision variables: where are the mines?
  Mines = new_array(NRows,NCols),
  Mines :: 0..1,

  foreach (I in 1..NRows, J in 1..NCols)

    % only check those cells that have hints
    if ground(Matrix[I,J]) then

      % The number of neighboring mines must equal Matrix[I,J].
      Matrix[I,J] #= sum([Mines[I+A,J+B] :
                                A in -1..1, B in -1..1,
                                I+A >  0, J+B >  0,
                                I+A =< NRows, J+B =< NCols]),

      % If there is a hint in a cell, then it cannot be a mine.
      Mines[I,J] #= 0
    end
  end,
  solve(Mines),
  println(Mines).

problem(P) =>
  P = {{_,_,2,_,3,_},
       {2,_,_,_,_,_},
       {_,_,2,4,_,3},
       {1,_,3,4,_,_},
       {_,_,_,_,_,3},
       {_,3,_,3,_,_}}.
```

Fig. 2.2 A program for Minesweeper

Figure 2.2 gives a program for the problem. The predicate problem/1 specifies
an instance matrix in which an integer indicates a hint value and an underscore _
indicates an unknown. For an instance matrix, let NRows be the number of rows and
NCols be the number of columns. The function new_array(NRows,NCols)
creates a matrix of the same dimension as the instance matrix. The domain constraint
Mines :: 0..1 restricts every variable to be Boolean: the value 0 indicates that the
cell does not contain a mine, and the value 1 indicates that the cell contains a mine.
The loop generates constraints to ensure that all of the hint values are respected. For
each cell, let I be its row number and J be its column number. If Matrix[I,J]
is ground, meaning that the cell has a hint value, then Mines[I,J] must be 0, and
the sum of the neighboring cells in the matrix Mines must be equal to the hint value.

Table 2.1 Minesweeper: CP vs SAT

N	Solutions	CP (s)	SAT (s)	Winner
50	1	0.016	0.072	CP
107	1	0.068	0.208	CP
202	1	0.284	0.548	CP
430	1	1.96	2.19	CP
440	2	3.6	3.08	SAT
450	5	10.1	6.46	SAT
500	3	7.5	5.03	SAT
601	1	4.99	3.65	SAT
1000	1	21.69	9.47	SAT
1500	3	804.6	61.27	SAT
1601	1	143.25	40.79	SAT

Note how this constraint is expressed nicely using a list comprehension. For a cell whose position is [I,J], a neighboring cell must have the position [I+A,J+B], where A and B are row and column deltas in -1..1. In the list comprehension, conditions are tested to ensure that I+A is a valid row number and J+B is a valid column number.

This program uses the sat module. For large instances of the Minesweeper problem, the sat module tends to be significantly faster than other solver modules. In general, SAT solvers tends to outperform CP solvers on 0/1 integer programming modules. SAT solvers perform better because the search strategies that CP solvers use for this type of problem are limited, and because SAT solvers are much more intelligent than CP solvers in pruning unfruitful paths from the search space.

The authors did an experiment of generating 20 $N \times N$ random matrices with N random hint values in the range of 1 to 8, including both solvable and unsolvable Minesweeper instances. The timings for selected N with at least one solvable instance are shown in Table 2.1. This suggests that, for this setup, the CP solver is faster than the SAT solver for $N \leq 430$, and that, for larger N, the SAT solver is significantly faster. Note: The exact times are for Picat version 1.3, and might differ in later versions.

2.5 Diet—Mathematical Modeling with the mip Module

This section considers the *Diet problem*, a popular linear programming problem in Operations Research. Given a set of foods, each of which has given nutrient values, a cost per serving, and a minimum limit for each nutrient, the objective of the diet problem is to select the number of servings of each food to consume so as to minimize the cost of the food while meeting the nutritional constraints. Table 2.2 gives an example. In this example, a diet is required to contain at least 500 calories, 6 ounces of chocolate, 10 ounces of sugar, and 8 ounces of fat.

Table 2.2 An example of the diet problem

Type of food	Calories	Chocolate (oz.)	Sugar (oz.)	Fat (oz.)	Price (cents)
Chocolate cake (1 slice)	400	3	2	2	50
Chocolate ice cream (1 scoop)	200	2	2	4	20
Cola (1 bottle)	150	0	4	1	30
Pineapple cheesecake (1 piece)	500	0	4	5	80
Limits	500	6	10	8	–

```
import mip.

main =>
  data(Prices,Limits,{Calories,Chocolate,Sugar,Fat}),
  Len = length(Prices),
  Xs = new_array(Len),
  Xs :: 0..10,

  scalar_product(Calories,Xs, #>=,Limits[1]), % 500
  scalar_product(Chocolate,Xs,#>=,Limits[2]), % 6
  scalar_product(Sugar,Xs,    #>=,Limits[3]), % 10
  scalar_product(Fat,Xs,      #>=,Limits[4]), % 8
  scalar_product(Prices,Xs,XSum), % to minimize
  solve($[min(XSum)],Xs),
  writeln(Xs).

% plain scalar product
scalar_product(A,Xs,Product) =>
  Product #= sum([A[I]*Xs[I] : I in 1..A.length]).

scalar_product(A,Xs,Rel,Product) =>
  scalar_product(A,Xs,P),
  call(Rel,P,Product).

data(Prices,Limits,Nutrition) =>
  % prices in cents for each product
  Prices = {50,20,30,80},
  % required amount for each nutrition type
  Limits = {500,6,10,8},

  % nutrition for each product
  Nutrition =
    {{400,200,150,500},  % calories
     {  3,  2,  0,  0},  % chocolate
     {  2,  2,  4,  4},  % sugar
     {  2,  4,  1,  5}}. % fat
```

Fig. 2.3 A program for the diet problem

Figure 2.3 gives a program for the example problem. The predicate

```
data(Prices,Limits,Nutrition}
```

defines a problem instance, where `Prices` is an array of prices of the foods, `Limits` is an array of the minimum numbers of servings for the different nutrients, and `Nutrition` is a matrix that gives the nutrient values of each type of food. The array `Xs` is a vector of decision variables. For each food type numbered `I`, `Xs[I]` denotes the number of servings of the food. The maximum number of servings is set to 10.

This program uses two user-defined constraints, named `scalar_product/3` and `scalar_product/4`, to handle the nutritional constraints. Let `A` be a vector of nutrient values. The `scalar_product(A,Xs,Product)` constraint ensures that `Product` is the total value of the nutrient. The `scalar_product/4` constraints ensure that the total value of each nutrient is no less than the required limit.

The call `scalar_product(Prices,Xs,XSum)` creates a new decision variable, `XSum`, which is equal to the total price of the foods. This is an optimization problem; the option in `solve($[min(XSum)],Xs)` means that `XSum` is minimized.

The program can be shortened by using a loop. The `main` predicate can be rewritten into the following:

```
main =>
  data(Prices,Limits,Nutrients),
  Len = length(Prices),
  Xs = new_array(Len),
  Xs :: 0..10,
  foreach (I in 1..Nutrients.length)
    scalar_product(Nutrients[I],Xs,#>=,Limits[I])
  end,
  scalar_product(Prices,Xs,XSum),
  solve($[min(XSum)],Xs),
  writeln(Xs).
```

This shortened program is more general than the previous version, and works on an input table of any size.

Modeling tip: For beginners, it might be difficult to start by implementing a very general model. Therefore, it is often beneficial to first write a small "explicit" model, which can then be generalized. Also, when writing a new model, it is recommended to start with a small problem instance, where the solutions are known. This has the advantages of being faster to solve, and that one can directly check that the solver has given a correct solution.

In this example, when modeling the diet problem, all three of the constraint modules quickly solve the problem. The next section gives an example for which `mip` is significantly faster than `sat` and `cp`.

2.6 The Coins Grid Problem: Comparing MIP with CP/SAT

As indicated earlier, the same Picat model can often be used for all three solver modules. This section studies the *Coins Grid problem*, an example on which the MIP solver is much faster than the other two solvers.

The problem statement is from Tony Hürlimann's "A coin puzzle: SVOR-contest 2007",[2] and is described below:

> In a quadratic grid (or a larger chessboard) with 31×31 cells, one should place coins in such a way that the following conditions are fulfilled:
>
> 1. In each row exactly 14 coins must be placed.
> 2. In each column exactly 14 coins must be placed.
> 3. The sum of the quadratic horizontal distance from the main diagonal of all cells containing a coin must be as small as possible.
> 4. In each cell at most one coin can be placed.

Figure 2.4 gives a program for the problem. The grid is represented as a matrix with 0/1 entries, with 0 indicating that a coin is not placed in the cell, and 1 indicating that a coin is placed in the cell. The row and column constraints are easy to model using `sum`. For a cell in row `I` and column `J`, the "quadratic horizontal distance" of the cell from the main diagonal is `abs(I-J)*abs(I-J)`. Sum is the total of the distances between the coins and the grid's main diagonal. The objective of this problem is to minimize `Sum`.

Table 2.3 compares the three solvers using the same model on different grid sizes (`N`) and different numbers of coins (`C`). The MIP solver solves the original problem (`N=31, C=14`) in a few milliseconds, whereas both CP and SAT failed to solve this problem in reasonable times (several hours). When compared to the MIP solver, both the CP and SAT solvers are very slow, even for small values of `N` and `C`.

The reason why the MIP solver is much faster is probably that all of the constraints in the model are linear, and MIP solvers are often very fast for linear models.

Modeling tip: When working with harder optimization problems, it is often useful to see the current value of the objective variable during the solving phase. In Picat, the option `report` (*Call*) can be used for this purpose. For example,

```
solve($[min(Sum),report(printf("Sum: %w\n", Sum))],Vars),
```

tells Picat to print out "Sum: " whenever it finds a new and better value of `Sum`.

---

[2]Taken with permission from www.svor.ch/competitions/competition2007/AsroContestSolution.pdf.

```
import mip.

main =>
  N = 31,
  C = 14,
  time2(coins(N, C)).

coins(N,C) =>
  X = new_array(N,N),
  X :: 0..1,

  foreach (I in 1..N)
    C #= sum([X[I,J] : J in 1..N]), % rows
    C #= sum([X[J,I] : J in 1..N])  % columns
  end,

  % quadratic horizontal distance
  Sum #= sum([(X[I,J] * abs(I-J) * abs(I-J)) :
              I in 1..N, J in 1..N]),
  solve($[min(Sum)],X),
  println(sum=Sum).
```

Fig. 2.4 A program for the coins-grid problem

Table 2.3 Coins grid problem (Picat 1.3)

N	C	Sum	CP (s)	SAT (s)	MIP (s)
8	4	80	2.27	3.48	0.0
8	5	198	72.15	3.76	0.0
9	4	90	12.61	6.48	0.0
10	4	98	97.39	7.51	0.0
31	14	13668	–	–	0.016

2.7 *N*-Queens—Different Modeling Approaches

The *N-queens problem* is another standard CSP problem. The objective is to place *N* queens on an *N* × *N* chessboard, such that no queen can capture any other queen. Queens on a chessboard can go in three directions: horizontally, vertically, and diagonally. There are several ways of modeling this problem. This section presents three models.

Here is the first, "naive", model:

```
import cp.

queens1(N, Q) =>
  Q = new_list(N),
  Q :: 1..N,
```

```
foreach (I in 1..N, J in 1..N, I < J)
  Q[I] #!= Q[J],          % columns
  Q[I] - I #!= Q[J] - J, % diagonal 1
  Q[I] + I #!= Q[J] + J  % diagonal 2
end,
solve([ff],Q).
```

This model uses a list Q of N variables, representing the chessboard's *N* rows, where each variable's value is a column number for each row. For all different pairs of I and J (where I < J), this model ensures the following:

1. The columns are distinct: Q[I] #!= Q[J].
2. No two queens are placed the same diagonal: Q[I] - I #!= Q[J] - J, and Q[I] + I #!= Q[J] + J.

The second model uses the global constraint all_different/1 for handling the three principal constraints:

```
import cp.

queens2(N, Q) =>
  Q = new_list(N),
  Q :: 1..N,
  all_different(Q),
  all_different([$Q[I]-I : I in 1..N]),
  all_different([$Q[I]+I : I in 1..N]),
  solve([ff],Q).
```

As shown in earlier examples, the arguments of the latter two all_different constraints must be "escaped" by $ in order for them to be treated as terms rather than as function calls.

Recall that the option [ff] specifies the *first-fail* labeling strategy. More about labeling strategies will discussed in the next chapter.

A third model is a 0/1 integer programming model that uses an $N \times N$ array. For each cell, a 0/1 domain variable is used, with 0 indicating an empty square, and 1 indicating a queen. Figure 2.5 shows a program based on this model.

The first two foreach loops ensure that there is exactly one queen in each row and each column. The two latter loops ensure the diagonality constraints by restricting each possible diagonal in the grid to contain at most one queen. Note that some diagonals might not contain any queens.

The timings of the three *N*-queens versions for the three constraint solvers, shown in Tables 2.4, 2.5 and 2.6, indicate some interesting differences. For the CP solver, the all_different version is significantly faster than both the SAT and MIP solvers. For the SAT solver, the "naive" version is faster than the all_different version, which perhaps indicates that, for this problem, the translation of the constraint all_different to the SAT solver's representation is not as good as

```
import sat.

queens3(N, Q) =>
  Q = new_array(N,N),
  Q :: 0..1,
  foreach (I in 1..N)
    % 1 in each row
    sum([Q[I,J] : J in 1..N]) #= 1
  end,
  foreach (J in 1..N)
    % 1 in each column
    sum([Q[I,J] : I in 1..N]) #= 1
  end,
  foreach (K in 1-N..N-1)
    % at most 1 in each \ diagonal
    sum([Q[I,J] : I in 1..N, J in 1..N, I-J==K]) #=< 1
  end,
  foreach (K in 2..2*N)
    % at most 1 in each / diagonal
    sum([Q[I,J] :  I in 1..N, J in 1..N, I+J==K]) #=< 1
  end,
  solve(Q).
```

Fig. 2.5 A 0/1 integer programming model for N-queens

Table 2.4 N-queens: "Naive version", using solve($[ff],Q) for CP

N	CP (s)	SAT (s)	MIP (s)
8	0.00	0.08	0.35
10	0.00	0.12	0.50
12	0.00	0.16	5.17
20	0.00	0.80	1557.3
50	0.00	8.74	–
100	0.02	48.10	–
200	4.86	105.61	–
400	0.56	–	–
1000	5.02	–	–

the implemented "naive" version. Note that the MIP solver does not support the all_different constraint. For the third model, the SAT solver is by far the fastest among the three solvers. Also, this is the model where the MIP solver has its fastest times.

Comparing all the models and solvers, the fastest combination is the CP solver running the all_different model.

Table 2.5 *N*-queens: all_different version, using solve($[ff],Q) for CP

N	CP (s)	SAT (s)	MIP (s)
8	0.00	0.09	–
10	0.00	0.14	–
12	0.00	0.22	–
20	0.00	0.92	–
50	0.004	8.33	–
100	0.01	473.1	–
200	2.36	782.88	–
400	0.28	–	–
1000	1.8	–	–

Table 2.6 *N*-queens: 0/1 integer programming version, using solve($[inout],Q) for CP

N	CP (s)	SAT (s)	MIP (s)
8	0.0	0.0	0.0
10	0.0	0.0	0.0
12	0.0	0.0	0.02
20	0.0	0.05	0.27
50	0.05	0.93	12.38
100	10.17	1.65	704.7
200	>60 min	7.20	–
400	–	48.41	–
1000	–	704.11	–

The timings are for Picat version 1.3 using Linux Ubuntu 12.04, 64-bit, Intel i7-930 CPU (2.8 GHz), 12 Gb RAM. Later versions might give different timings (and even different relations between the solvers' times).

As indicated in this chapter, testing different approaches and solvers can give very different behaviors in terms of solving times. Sometimes, this kind of experimentation is not really needed, since the first variant is "good enough". Nevertheless, trying different variants is a good way of learning constraint modeling, and can also be fun.

2.8 Bibliographical Note

This chapter and Chap. 3 cover a few techniques and applications of constraint programming. A more in-depth discussion of constraint programming, including its history, the algorithms that constraint solvers use, and further techniques and applications of constraint programming, is included in [45].

The literature on constraint programming is abundant, including tutorials [3, 7], surveys [10, 28, 44]) and textbooks ([18, 37, 57, 58]). Picat's CP solver is inherited from B-Prolog [67, 68].

The Diet problem is just one type of Operations Research problem. References [54] and [65] contain examples of how Operations Research can be used to manage businesses.

The description of the Coins Grid problem is taken from [27].

2.9 Exercises

1. Implement the *carry version* of the SEND + MORE = MONEY problem from Sect. 2.2. In the carry version, whenever the the sum of the column is greater than ten, the leftmost digit of the sum is explicitly added to the next column on the left. For example,

```
  C4  C3  C2  C1
        S   E   N   D
  +     M   O   R   E
  ---------------
  = M   O   N   E   Y
```

The four variables C1, C2, C3, and C4 are the carries in the addition.
Compare the performance of the two models.

2. Implement the DONALD + ROBERT = GERALD problem using both the non-carry approach and the carry approach. Compare the performances.

3. Modify the Sudoku program from Sect. 2.3.1 to solve the following 9×9 Sudoku instance:

```
_ _ 2   _ _ 5   _ 7 9
1 _ 5   _ _ 3   _ _ _
_ _ _   _ _ _   6 _ _

_ 1 _   4 _ _   9 _ _
_ 9 _   _ _ _   _ 8 _
_ _ 4   _ _ 9   _ 1 _

_ _ 9   _ _ _   _ _ _
_ _ _   1 _ _   3 _ 6
6 8 _   3 _ _   4 _ _
```

Ensure that it has only one solution.

4. The output of the `minesweeper` model in Sect. 2.4 is just a list of 0s and 1s.
   Add a predicate to print the output in the following format:
   The mine cells are represented with X, and the non-mine cells are shown as _.
   Check that the minesweeper solution complies with the given hints.
5. Write a constraint model for the *Zebra problem* (sometimes attributed to Lewis
   Carroll):

   > Five men with different nationalities live in the first five houses of a street. They practice
   > five distinct professions, and each of them has a favorite animal and a favorite drink,
   > all of them different. The five houses are painted in different colors.
   > The Englishman lives in a red house.
   > The Spaniard owns a dog.
   > The Japanese is a painter.
   > The Italian drinks tea.
   > The Norwegian lives in the first house on the left.
   > The owner of the green house drinks coffee.
   > The green house is on the right of the white one.
   > The sculptor breeds snails.
   > The diplomat lives in the yellow house.
   > Milk is drunk in the middle house.
   > The Norwegian's house is next to the blue one.
   > The violinist drinks fruit juice.
   > The fox is in a house next to that of the doctor.
   > The horse is in a house next to that of the diplomat.
   >
   > Who owns a zebra, and who drinks water?

   Ensure that the model has a unique solution.
6. *Map coloring*:

   (a) The map coloring problem is to color the countries of a map, ensuring that
   each country is given a different color than the countries that are its neighbors.
   Below are some European countries and their neighbors. Implement a map coloring model for these countries.

   > Countries: Belgium, Denmark, France, Germany, Luxembourg, Netherlands
   >
   > Neighbors:
   >
   > - Belgium: France, Germany, Luxembourg, Netherlands
   > - Denmark: Germany
   > - France: Belgium, Germany, Luxembourg
   > - Germany: Belgium, Denmark, France, Netherlands, Luxembourg
   > - Luxembourg: Belgium, France, Germany
   > - Netherlands: Belgium, Germany

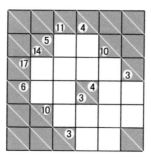

Fig. 2.6 Sample Kakuro puzzle

Print all the possible solutions.

Hint 1: The colors themselves are not very important, and can be represented as integers.

Hint 2: There are many solutions which are the same, except that the numbers are a permutation of another solution. To remove some of these *symmetries* one can assume, for example, that Belgium has the color 1.

(b) Find a map of Europe and implement a map coloring model of all the European countries.

7. Test the models in this chapter with the cp and sat constraint modules and compare the results.

8. Solve the instance of the *Kakuro puzzle* that is depicted in Fig. 2.6. The objective of the Kakuro puzzle is to fill in the white squares with digits between 1 and 9 such that each set of horizontal digits adds up to the value on its left, and each set of vertical digits adds up to the value above it. No digit can be used more than once in each sum. Note: when one square contains two pre-filled numbers, the top number applies to the horizontal sum, and the bottom number applies to the vertical sum.[3]

---

[3]This is a relatively small problem instance of Kakuro. For larger and more difficult instances, see the website Kakuro Conquest (http://www.kakuroconquest.com/).

Chapter 3
Advanced Constraint Modeling

Abstract This chapter describes more advanced techniques of constraint modeling in Picat, as well as some tips about performance and debugging. The *Langford number problem* shows the use of the global constraint `element` and the concepts of *reversibility* and *symmetry breaking*. *Reification*, or reasoning about constraints, is exemplified with a decomposition of `all_different_except_0`, as well as the *Who Killed Agatha* problem. A separate section describes the importance of declaring domains to be as small as possible (but not smaller). This is followed by a section about search strategies, in which the *magic squares problem* is used for systematically testing many *search strategies*. The `cumulative` constraint is used for *scheduling* furniture moving, including *precedences*. Then, the *magic sequences problem* shows how to use the global constraint `global_cardinality`, and explains that the order of constraints in a model might matter. The `circuit` constraint is used for modeling the *knight's tour problem*. The `regular` constraint, defined by a *Deterministic Finite Automaton*, is used in two models, first for a decomposition of the `global_contiguity` constraint, and then for modeling a *nurse rostering* problem. This is followed by an alternative approach of the nurse rostering problem that uses the `table_in` constraint. The chapter ends with some general tips about how to debug constraint models, when the models don't give a solution, or when the solution is not correct.

3.1 Advanced Constraint Modeling

Chapter 2 described basic modeling with the three constraint modules: `cp`, `sat`, and `mip`. This chapter describes more advanced modeling with a focus on global constraints and the `cp` and `sat` modules.

This chapter also describes some of the built-in *global constraints*, including `all_different`, `element`, `lex_lt`, `lex_le`, `all_different_except _0`, `cumulative`, `global_cardinality`, `scalar_product`, `circuit`, `regular`, `table_in`, and `table_not_in`. See the Picat Guide for a complete list of the global constraints that Picat provides. For a more comprehensive list of

© The Author(s) 2015
N.-F. Zhou et al., *Constraint Solving and Planning with Picat*,
SpringerBriefs in Intelligent Systems, DOI 10.1007/978-3-319-25883-6_3

global constraints, see The Global Constraint Catalog.[1] For a collection of constraint satisfaction problems, some of which are described in this chapter, see CSPLib.[2]

Note: The `mip` module in Picat version 1.3 does not support the global constraints that are described in this chapter.

Note: Some of the models using the `sat` solver do not work in the Windows version of Picat, because the Lingeling SAT solver doesn't compile with MVC. Windows users can install Cygwin, and use the Cygwin version instead.

3.2 The `element` Constraint and Langford's Number Problem

This section introduces some of the important concepts of constraint modeling:

- the `element` constraint
- *reversibility*
- *symmetry breaking*

The `element` constraint[3] is a very important and commonly-used constraint. In some modeling languages, such as MiniZinc and OPL, it is denoted as V = L[I], where L is a list of decision variables or integers, and each of I and V can either be a decision variable or an integer. The `element` constraint holds if V is the value of the *I*th position in the list L. However, Picat does not support the V #= L[I] syntax when the index, I, is a decision variable. Instead, the constraint must be stated as `element(I,L,V)`, and should be interpreted as "V is the value of L at index I".

If I and V are both decision variables, and if L is a list of decision variables or integers, then the constraint can be used in two different cases:

- It can be used for the "normal" index operation that is found in many programming languages: Given the value of I, which is the index in L, calculate the value of V.
- It can also be used for the reverse operation: Given the value of V (the result value), figure out what index I must be.

These multiple ways of using a constraint are an example of *reversibility* (also called *bi-directionality*), where certain parameters of the constraint can either be used as input or as output, depending on the circumstances. In the first case, I and L are inputs, and V is the output. In the second case, L and V are inputs, and I is the output.

---

[1] http://sofdem.github.io/gccat/.

[2] http://csplib.org/.

[3] http://sofdem.github.io/gccat/gccat/Celement.html.

```
                    import cp.
                    main =>
                      langford(4, Seq),
                      println(Seq).

                    langford(K, Seq) =>
                      K2 = 2*K,
                      Seq = new_list(K2),
                      Seq :: 1..K,
                      foreach(I in 1..K)
                          element(J1,Seq,I),
                          element(J2,Seq,I),
                          J2 #= J1+I+1
                      end,
                      solve(Seq).
```

Fig. 3.1 Langford problem, first approach

3.2.1 Langford's Number Problem

The usefulness of the element constraint can be shown by the *Langford number problem*.[4]

Consider two sets of the numbers from 1 to 4. The problem is to arrange the eight numbers in the two sets into a single sequence in which the two 1's appear one number apart, the two 2's appear two numbers apart, the two 3's appear three numbers apart, and the two 4's appear four numbers apart.

For example, for $k = 4$, a solution would be 41312432.

This section contains two models of the Langford number problem. The first model, shown in Fig. 3.1, uses a single list, Seq, with the domain 1..K, and two element constraints. The loop checks that for each number 1..K, the two indices at which I is placed in Seq, indices J1 and J2, are I + 1 positions apart. In other words, the loop ensures that there are exactly I numbers between the two occurrences of I in Seq.

The output for K = 4 is this solution:

```
[4,1,3,1,2,4,3,2]
```

The advantages of this model are that it is small, and that it is straightforward to implement. However, the drawback is that it is slow when given larger values of K. For example, for as small a size as K = 8, it takes some seconds to solve, and K = 12 takes minutes.

Another approach is to add another *viewpoint* in the model: a second list that also contains the *positions* of the numbers. This approach is shown in Fig. 3.2.

The two lists Solution and Position are of length 2*K in order to contain the double set of the integers 1..K, where the two occurrences of the digit I should

[4]Taken with permission from http://www.csplib.org/Problems/prob024/.

```
import cp.

main =>
  langford(4, Solution, Position),
  println(solution=Solution),
  println(position=Position).

langford(K, Solution, Position) =>
  K2 = 2*K,
  Position = new_list(K2), Position :: 1..K2,
  Solution = new_list(K2), Solution :: 1..K,

  all_distinct(Position),

  foreach (I in 1..K)
     Position[K+I] #= Position[I] + I+1,
     element(Position[I],Solution,I),
     element(Position[K+I],Solution,I)
  end,

  % symmetry breaking:
  Solution[1] #< Solution[K2],

  Vars = Solution ++ Position,
  solve(Vars).
```

Fig. 3.2 Langford problem, second approach

be placed I places apart. Position contains the *position* of the integers, where the same integer is placed in positions I and K+I. For example, for K = 4 the position of the number 3 will be stored in Position[3] and Position[4+3]. For K = 4, one solution is Solution=41312432, which means that Position is 25314876.

The two element constraints convert the positions in the list Position to the solution list, Solution, by ensuring that the index of the number I (i.e., the index that is stored in Position[I]) will be the number I.

The advantage of this second approach is that the "viewpoint" of the Solution list makes it possible to add an all_different constraint, which speeds up the model considerably. Comparing with the first model, an input of K = 12 is solved in 0.004 s, and all K up to 32 are solved in well under 1 s.

A short comment on *symmetry breaking*: For K = 4 there are two solutions:

```
solution:[4,1,3,1,2,4,3,2]
position:[2,5,3,1,4,8,7,6]
```

and

```
solution:[2,3,4,2,1,3,1,4]
position:[5,1,2,3,7,4,6,8]
```

These two solutions are the same, except that one is the reverse of the other. When distinct solutions are required, removing these kind of symmetries are done with various types of *symmetry breaking* constraints.

One way to break the symmetry in this model is to ensure that the first element in `Solution` is less than the last element:

```
Solution[1] #< Solution[K2]
```

Adding this symmetry breaking constraint will only give the second solution.

Using symmetry breaking can make a model much faster, since it removes many possible solutions by pruning branches in the search tree. There are other global constraints designed to be used for symmetry breaking. Picat supports the two constraints `lex_lt(L1,L2)` and `lex_le(L1,L2)` for ensuring that the list `L1` is either lexicographically *less than* (`lex_lt`) or *less than or equal to* (`lex_le`) the list `L2`. Another common symmetry breaking constraint is `lex2(Matrix)`, which ensures that all rows and columns in a 2-dimensional matrix are ordered lexicographically. Note that `lex2` is not a Picat built-in.

3.3 Reification

Reification is an invaluable feature of constraint modeling, and makes it possible to reason about constraints by stating logical constraints, and by combining them with logical operators.

Table 3.1 defines the special operators that can be used when working with reifications.

Table 3.1 Reification operators in Picat

Operation	Operator	Definition
and	`C1 # /\ C2`	Constraint `C1` and constraint `C2` must both be true
or	`C1 #\ / C2`	Either constraint `C1` or constraint `C2` (or both) must be true
inequality	`C1 #!= C2`	If constraint `C1` is true, then constraint `C2` is false
		If constraint `C1` is false, then constraint `C2` is true
implication	`C1 #=> C2`	If constraint `C1` is true, then `C2` must be true
equivalence	`C1 #<=> C2`	Both constraints `C1` and `C2` must be true
		or both constraints `C1` and `C2` must be false
xor	`C1 #^ C2`	Either constraint `C1` or constraint `C2`, but not both, must be true
not	`#~ C1`	Constraint `C1` is not true

3.3.1 Example of Reification :
all_different_except_0

The global constraint `all_different_except_0(Xs)`[5] is a useful example
of reification. This constraint ensures that all of the non-zero values in the list of
decision variables `Xs` must be different. In other words, the number 0 can appear
multiple times in list `Xs`, but other values can appear at most once.

Even though Picat supports this constraint, it is instructive to see how it can be
defined using reifications (Fig. 3.3).

The constraint loops through every pair of elements in `Xs` (`X[I]` and `X[J]`,
`I > J`), and uses implication to ensure that when both `Xs[I]` and `Xs[J]` are
different from 0, they are distinct.

This constraint is useful when the list can contain "dummy" values (coded as the
value 0) which should be ignored when ensuring the overall distinctiveness of the
elements in a list. Note that the domain of the list of decision variables must include
the value 0.

3.3.2 Reification—Who Killed Agatha

A more elaborate use of reification is used in the *Who Killed Agatha* problem, which
is also called "The Dreadsbury Mansion Murdery Mystery." This is a standard benchmark for theorem proving. It is stated like this:

> Someone in Dreadsbury Mansion killed Aunt Agatha. Agatha, the butler, and Charles live
> in Dreadsbury Mansion, and are the only ones to live there. A killer always hates, and is
> no richer than his victim. Charles hates nobody that Agatha hates. Agatha hates everybody

```
import cp.

main =>
  N = 4,
  Xs = new_list(N),
  Xs :: 0..N,
  alldifferent_except_0(Xs),
  solve(Xs),
  println(Xs).

alldifferent_except_0(Xs) =>
  foreach (I in 2..Xs.length, J in 1..I-1)
     (Xs[I] #!= 0 #/\ Xs[J] #!= 0) #=> (Xs[I] #!= Xs[J])
  end.
```

Fig. 3.3 All different except 0

[5]http://sofdem.github.io/gccat/gccat/Calldifferent_except_0.html.

except the butler. The butler hates everyone not richer than Aunt Agatha. The butler hates everyone whom Agatha hates. Nobody hates everyone. Who killed Agatha?

The Picat model below shows some examples of reifications, but also demonstrates caveats when modeling a problem: Sometimes, modeling a problem involves more than just translating constraints from a problem description. One also has to define the underlying *concepts*. In this example, the underlying concepts are the concepts of "Hate" and "Richer". This example also shows that sometimes the problem can be underdefined: This model yields eight solutions, all pointing to the same killer, since there are some insufficiently defined relations regarding "Hate" and "Richer" that cannot be decided from the problem description.

This model has two central concepts: "Hate" and "Richer". They are defined by two 3 × 3 matrices of Boolean decision variables. Reification is used to define both the principle of the concepts and the concrete relations from the problem description.

"A killer always hates, and is no richer than his victim."
One way of stating this constraint is to use a foreach loop which iterates through each person and states that "if this person is the killer, then he/she hates Agatha and he/she is not richer than Agatha":

```
foreach (I in 1..N)
   Killer #= I #=> Hates[I, Agatha] #= 1,
   Killer #= I #=> Richer[I, Agatha] #= 0
end,
```

The next step is to define the concept of "Richer" so the solver can use the given hints to identify the killer. In this model, to be richer means that:

- nobody is richer than him-/herself
- if person P1 is richer than person P2, then P2 cannot be richer than P1

The second criterion uses a reification constraint (an equivalence):

```
% if I is richer than J, then J is not richer than I
foreach (I in 1..N, J in 1..N, I != J)
   Richer[I,J] #= 1 #<=> Richer[J,I] #= 0
end,
```

Similarly, the following requirements also use reification constraints:

- Charles hates nobody that Agatha hates.
- The butler hates everyone not richer than Aunt Agatha.
- The butler hates everyone whom Agatha hates.

The output of the model is a list of eight 1's, stating that, in all eight possible solutions, Agatha, who is labeled as person 1, is the killer (Fig. 3.4).

```
import cp.

main =>
    L = find_all(Killer, who_killed_agatha(Killer)),
    println(killer=L).

who_killed_agatha(Killer) =>
    % Agatha, the butler, and Charles live in Dreadsbury Mansion,
    % and are the only ones to live there.
    N = 3,
    Agatha = 1,
    Butler = 2,
    Charles = 3,

    % The killer is one of the three.
    Killer :: [Agatha,Butler,Charles],
    % Define the Hates and Richer Boolean matrices
    Hates = new_array(N,N),
    Hates :: 0..1,
    Richer = new_array(N,N),
    Richer :: 0..1,

    % A killer always hates, and is no richer than his victim.
    foreach (I in 1..N)
        Killer #= I #=> Hates[I, Agatha] #= 1,
        Killer #= I #=> Richer[I, Agatha] #= 0
    end,
    % Define the concept of "richer":
    % nobody is richer than him-/herself
    foreach (I in 1..N)
        Richer[I,I] #= 0
    end,
    % if I is richer than J, then J is not richer than I
    foreach (I in 1..N, J in 1..N, I != J)
        Richer[I,J] #= 1 #<=> Richer[J,I] #= 0
    end,
    % Charles hates nobody that Agatha hates.
    foreach (I in 1..N)
        Hates[Agatha, I] #= 1 #=> Hates[Charles, I] #= 0
    end,
    % Agatha hates everybody except the butler.
    Hates[Agatha, Butler]  #= 0,
    Hates[Agatha, Charles] #= 1,
    Hates[Agatha, Agatha]  #= 1,
    % The butler hates everyone not richer than Aunt Agatha.
    foreach (I in 1 ..N)
        Richer[I, Agatha] #= 0 #=> Hates[Butler, I] #= 1
    end,
    % The butler hates everyone whom Agatha hates.
    foreach (I in 1..N)
        Hates[Agatha, I] #= 1 #=> Hates[Butler, I] #= 1
    end,
    % Nobody hates everyone.
    foreach (I in 1..N)
        sum([Hates[I,J] : J in 1..N]) #=< 2
    end,
    % Who killed Agatha?
    Vars = [Killer] ++ Hates ++ Richer,
    solve(Vars).
```

Fig. 3.4 "Who killed Agatha", using *reification*

3.4 Domains: As Small as Possible (but Not Smaller)

The underlying algorithms of the constraints in the cp module are based on the reduction of the domains of the decision variables, either by adjusting the lower and upper bounds, or by deleting values from the domain.

For this reason, it is important to declare the domains as small as possible, in order to avoid any unnecessary work that the solver has to do in order to deal with irrelevant values, such as values that are too small or too large.

Sometimes, during the reading phase, the solver can figure out the lower and upper bounds of a domain directly. For example, when using sum:

```
% ...
N = 10,
Xs = new_list(N),
Xs :: 1..N,

% ...
Z #= sum(Xs),
solve(X).
```

The cp module will figure out while reading the constraint that the smallest possible value of Z is 10 (i.e., $10 * 1$) and that the largest possible value is 100 (i.e., $10 * 10$), so the domain of Z is 10..100. However, it is always good to make it a habit to declare the domains when possible.

When building and debugging a model, it is also good practice to print the decision variables to see whether the domains seem reasonable. The following is an example of printing decision variables:

```
Picat> import cp
Picat> Xs = new_list(3), Xs :: 1..10, Z #= sum(Xs)
Xs = [DV_013968_1..10,DV_0139b8_1..10,DV_013a08_1..10]
Z = DV_013ca0_3..30
```

Here, the result from the Picat interpreter is shown, to emphasize that this kind of testing can be done both in the model and in the interpreter for testing small code snippets. The output shows Xs as it is represented in Picat. A decision variable is shown as DV_Id_<Domain>, e.g., DV_013968_1..10, where "013968" is an internal id, and 1..10 is the domain of the decision variable.

Note that if a value has been removed from the domain, then it will not be shown when the domain is printed. For example, when a decision variable is constrained to be an even number between 1 and 10, it has the domain {2, 4, 6, 8, 10}, which is presented like this:

```
Picat> A :: 1..10, A mod 2 #= 0
A = DV_0111b8_2_4_6_8_10
```

As mentioned in Sect. 2.3.2, when Picat's cp module reads a constraint, it tries to reduce the domains as much as possible, based on the propagation rules of the constraint. In general, domain reduction alone cannot lead to a solution, and search is necessary to find solutions. Note that the sat module does not perform this type of domain reduction until solve is called.

3.5 Search and Search Strategies (cp Module)

As mentioned in Chap. 2, one of the ways to make a cp model more efficient is to use different search strategies, or *heuristics*. A heuristic is a strategy for deciding the order in which variables are selected and the order in which values are tried for the selected variable.

Picat has a number of different search strategies, including the following[6]:

Variable selection: These affect the order in which the decision variables are selected:

- backward: Before searching, the list of variables is reversed.
- constr: Variables are first ordered by the number of constraints to which they are attached.
- degree: Variables are first ordered by *degree*, i.e., the number of connected variables.
- ff: The *first-fail principle* is used: select the leftmost variable with the smallest domain.
- ffc: This is a combination of ff and constr.
- ffd: This is a combination of ff and degree.
- forward: Choose variables in the given order, from left to right.
- inout: The variables are reordered in an inside-out fashion. For example, the variable list [X1,X2,X3,X4,X5] is rearranged into the list [X3,X2, X4,X1,X5].
- leftmost: This is the same as forward.
- max: First, select a variable whose domain has the largest upper bound, breaking ties by selecting a variable with the smallest domain.
- min: First, select a variable whose domain has the smallest lower bound, breaking ties by selecting a variable with the smallest domain.
- rand_var: Variables are randomly selected when labeling.

Value selection: These strategies determine in which order values should be tested on the selected variable:

- down: Values are assigned to the variable from the largest value to the smallest value.
- rand_val: Values are randomly selected when labeling.

---

[6]This section only applies to the cp module. The sat module ignores the search strategies.

- `reverse_split`: Bisect the variable's domain, excluding the lower half first.
- `split`: Bisect the variable's domain, excluding the upper half first.
- `updown`: Values are assigned to the variable from the values that are nearest to the middle of the domain.

When no search strategy is specified, Picat uses the default strategy: `forward` and `down`. Another possible search strategy is the combined strategy `rand`, where both variables and values are randomly selected when labeling.

Unfortunately, there is no simple answer for which labeling strategy is the best for a specific model. Instead, experimentation must be applied.

Modeling tip: The most common labelings to test first are:

- `solve` without any labeling options
- `ff` combined with `down` or `split`

Sometimes, these labelings give a good enough result. However, when working with larger problems, it might be necessary to systematically test all the labeling combinations. As an example, the magic squares problem will be used.

3.5.1 Magic Squares—Systematically Testing All Labelings

The *magic squares problem*[7] is a classic CSP problem where the objective is to create a square numerical matrix in which every row, every column, and both diagonals sum to the same number. Figure 3.5 gives a Picat model for this problem. The model is straightforward, using the `all_different/1` constraint and the built-in `sum/1`. The sum is precalculated to $N*(N*N+1)/2$. When ensuring that all of the values in the `Square` matrix are distinct, `Square.vars()` is used to extract the decision variables from the matrix to a list.

The predicate `magic` has two extra parameters, `VarSel` and `ValSel`, for the selected variable and the value strategies, respectively.

Two helper predicates are used for testing. The `timeout(Goal, Timeout, Status)` predicate calls the goal `Goal` with a maximum of `Timeout` milliseconds. If the predicate times out, the `Status` will be `time_out`. If the predicate succeeds, the status will be `success`.

For a predicate `P`, the `time2(P)` predicate shows both the runtime of the predicate in seconds, and the number of *backtracks*. The number of backtracks represents the complexity of the search: it is the number of times the solver backtracks in the search tree to get the solution. A lower number of backtracks is better than a higher number. The ideal is 0 backtracks, which means that the solution has been found without any backtracking. Puzzles such as Sudoku usually have 0 backtracks when given the proper search heuristics.

---

[7]Taken with permission from http://www.csplib.org/Problems/prob019/.

```
import cp.

main =>
  selection(VarSels),
  choice(ValSels),
  Timeout = 10000, % Timeout in milliseconds
  N = 7, % size of problem
  foreach (VarSel in VarSels, ValSel in ValSels)
      time2(time_out(magic(N,_Square,VarSel,ValSel),
             Timeout,Status)),
      println([VarSel,ValSel,Status])
  end.

magic(N,Square,VarSel,ValSel) =>
  NN = N*N,
  Sum = N*(NN+1)//2,% magical sum

  Square = new_array(N,N),
  Square :: 1..NN,

  all_different(Square.vars()),

  foreach (I in 1..N)
    Sum #= sum([Square[I,J] : J in 1..N]), % rows
    Sum #= sum([Square[J,I] : J in 1..N])  % columns
  end,

  % diagonal sums
  Sum #= sum([Square[I,I] : I in 1..N]),
  Sum #= sum([Square[I,N-I+1] : I in 1..N]),

  solve([VarSel,ValSel], Square).

% Variable selection
selection(VarSels) =>
  VarSels = [backward,constr,degree,ff,ffc,ffd,
             forward,inout,leftmost,max,min,up].

% Value selection
choice(ValSels) =>
  ValSels = [down,reverse_split,split,up,updown].
```

Fig. 3.5 Magic square

Sometimes the number of backtracks is a more interesting value for comparing different search strategies—as well as different models of the same problem—than the solving time, since it provides a hint about the search tree's structure.

In Table 3.2, "to" indicates a timeout status, "s" indicates the number of seconds, and "bt" indicates the number of backtracks. The "Sum" column and row show the total number of successful searches in the adjacent columns and rows. The result indicates that the magic square program should use some of the ffc or ffd variable

Table 3.2 Magic square ($N = 7$), different labelings, timeout 10 s (s/bt)

var/val	down	split	reverse_split	up	updown	Sum
backward	to	to	to	to	0.2/71610	1
constr	to	to	to	to	0.4/164240	1
degree	to	to	to	to	to	0
ff	to	to	to	to	0.0/15726	1
ffc	0.2/75115	0.2/0	0.2/0	0.2/52383	0.2/71609	5
ffd	0.0/978	0.0/0	0.0/0	0.0/692	0.0/1139	5
forward	to	to	to	to	0.2/70356	1
inout	to	to	to	to	0.0/13750	1
leftmost	to	to	to	to	0.2/70356	1
max	to	to	to	to	to	0
min	to	to	to	to	to	0
Sum	2	2	2	2	8	

selections. Interestingly, the updown value selection, in which the values closest to the center are taken first, is very good when used together with most of the variable selections. This suggests that further experiments are needed, such as a longer timeout or a larger problem instance.

3.6 Scheduling—The cumulative Constraint

There are many types of *scheduling problems*. In Constraint Programming, there is a dedicated global constraint cumulative/4[8] for solving a certain variant of the scheduling problem.

The model in Fig. 3.6 shows a schedule for planning how to move furniture, in which four people have gathered to move different kinds of furniture:

- piano
- 4 chairs, named chair1, chair2, chair3, and chair4
- bed
- table
- TV
- 2 shelves, named shelf1 and shelf2

There are also some *precedence* constraints, where a certain piece of furniture must be moved before another piece of furniture:

- shelf1 must be moved before the bed
- shelf2 must be moved before the table

---

[8]http://sofdem.github.io/gccat/gccat/Ccumulative.html.

```
import cp, util.

main =>
   data(1,Data,MaxPeople,MaxTime,Precedences),
   schedule(Data,MaxPeople,MaxTime,Precedences).

schedule(Data,MaxPeople,MaxTime,Precedences) =>
   N = Data.length,
   [_Names,PeopleNeeded,Duration] = Data.transpose(),
   NameIxMap = new_map([Data[I,1]=I : I in 1..N]),
   % decision variables
   Start = new_list(N),
   Start :: 0..MaxTime,
   % end time of the tasks
   End = new_list(N),
   End :: 0..MaxTime,
   % number of people needed
   NumPeople :: 0..MaxPeople,

   % constraints
   cumulative(Start, Duration, PeopleNeeded, NumPeople),
   % connect start and end time for each task
   foreach (Task in 1..N)
      End[Task] #= Start[Task] + Duration[Task]
   end,
   MaxEnd #= max(End), % maximum End time
   % precedences: task P1 must be finished before task P2 starts
   foreach ([P1,P2] in Precedences)
      End[NameIxMap.get(P1)] #< Start[NameIxMap.get(P2)]
   end,
   % search
   Vars = End ++ Start ++ [NumPeople],
   solve($[ff,split,min(MaxEnd)],Vars),
   nl,
   println(numPeople=NumPeople),
   println(maxEnd=MaxEnd),
   println(start=Start),
   println(end=End).

data(1,Data,MaxPeople,MaxTime,Precedences) =>
           % name, people, time (in minutes)
     Data = [
             [piano , 3, 30],
             [chair1, 1, 10], % chairs
             [chair2, 1, 10],
             [chair3, 1, 10],
             [chair4, 1, 10],
             [bed   , 3, 20],
             [table , 2, 15],
             [shelf1, 2, 15], % shelves
             [shelf2, 2, 15],
             [tv    , 2, 15]
            ],
     MaxPeople = 4,
     MaxTime = sum([T : [_,_,T] in Data]),
     Precedences =
     [
       [shelf1,bed],
       [shelf2,table],
       [bed,piano],
       [tv,table]
     ].
```

Fig. 3.6 Furniture moving, using `cumulative`

- the bed must be moved before the piano
- the TV must be moved before the table

This example uses the global scheduling constraint

```
cumulative(Starts, Durations, Resources, Limit)
```

where `Starts` is a list of decision variables for each task (which refers to each piece of furniture in this example) with the domain `0..MaxTime`. The two lists `Durations` and `Resources` are lists with the durations and the number of people required for each task, respectively. In general, the `Starts`, `Durations`, and `Resources` lists can either be decision variables or lists of integers. `Limit` is the total number of resources needed to fulfill all the tasks, and can also be a decision variable or an integer. The `cumulative` constraint ensures that the limit cannot be exceeded at any given time.

Note that `cumulative/4` does not include the end times of the tasks. Therefore, it is necessary to add the `End` list in order to define the end time for each task as `End[Task] #= Start[Task] + Duration[Task]`. The decision variable `MaxEnd` is the largest end time, which is the total time required to perform all of the tasks.

The two common scenarios for scheduling these kinds of problems are to either minimize the maximum end time, or to minimize the number of resources needed (the `Limit` variable). This model uses the first scenario, minimizing the total time needed to move all of the furniture.

The precedence constraints are handled by ensuring that one task (`P1`) ends before the other one (`P2`) is started. This is accomplished by using the map `NameIxMap`. `NameIxMap` maps each furniture name to a unique integer which can be used as an index in the `Start` and `End` matrices. This allows the precedences to be stated within a foreach loop as

```
End[NameIxMap.get(P1)] #< Start[NameIxMap.get(P2)]
```

where `NameIxMap.get(P1)` extracts the integer that represents the name of the furniture `P1` from `NameIxMap`.

A note on data extraction: The model uses the `transpose` built-in from the `util` module to extract the three lists `Names`, `PeopleNeeded`, and `Duration` from the `Data` matrix:

```
import util.
   % ...
   [Names,PeopleNeeded,Duration] = Data.transpose()}.
   % ...
```

This is a more compact variant than using list comprehensions:

```
   % ...
   N = Data.length,
   Names          = [Data[I,1] : I in 1..N],
```

```
PeopleNeeded = [Data[I,2] : I in 1..N],
Duration]    = [Data[I,3] : I in 1..N],
% ...
```

3.7 Magic Sequence—`global_cardinality`/2 and the Order of Constraints

The *magic sequence problem*[9] is another classic CSP problem:

> A magic sequence of length n is a sequence of integers $x_0 .. x_{n-1}$ between 0 and n-1, such that for all i in 0 to n-1, the number i occurs exactly x_i times in the sequence. For instance, 6,2,1,0,0,0,1,0,0,0 is a magic sequence since 0 occurs 6 times in it, 1 occurs twice, ...

Figure 3.7 shows a model for the problem. The global constraint `global_cardinality(Sequence, Pairs)`[10] is a good fit for this problem, since its purpose is to handle the problem to count the occurrences of elements in `Sequence`. The decision variable list `Sequence` consists of a list of N elements with the domain of `0..N-1`, where the element `Sequence[I+1]` is the number of occurrences of the integer `I` in the list. Since Picat's lists and arrays are 1-based, the counts must be adjusted by adding 1 to each element's position in the `Sequence` list. `global_cardinality(Sequence, Pairs)` requires two arguments: the list of decision variables (which will be called `Sequence` in the magic sequence model), and a list of pairs of the cardinalities (which will be called `Pairs`). Each item in the `Pairs` list takes the form `K-V`, indicating that `K` must appear exactly `V` times in `Sequence`.

Using just the `global_cardinality` constraint is enough for solving the magic sequence problem, at least for smaller values. However, in order to speed up the model, some extra constraints are added. These constraints do not really influence the number of solutions (in contrast to "symmetry breaking" constraints); instead, they just make the model faster, and are therefore sometimes called *redundant constraints*. This model has two extra constraints based on observations about the nature of the problem:

- The sum of the numbers in the sequence must always be N.
- The scalar product of the numbers and the indices must be N. `scalar_product` is a built-in constraint.

Modeling tip: This model demonstrates an interesting property of Picat's cp module: The *order of constraints* in the model matters, and changes in the order can give an improvement of speed. If the `global_cardinality` is *before* the extra redundant constraint, then N = 400 takes about 8 s to solve. If, on the other hand,

---

[9]Taken with permission from http://www.csplib.org/Problems/prob019/.

[10]http://sofdem.github.io/gccat/gccat/Cglobal_cardinality.html.

```
import cp.

main =>
  magic_sequence(100,Sequence),
  println(Sequence).

magic_sequence(N, Sequence) =>
  Sequence = new_list(N),
  Sequence :: 0..N-1,

  % extra/redudant constraints to speed up the model
  N #= sum(Sequence),
  Integers = [I : I in 0..N-1],
  scalar_product(Integers, Sequence, N),

  % create list: [0-Sequence[1], 1-Sequence[2], ...]
  Pairs = [$I-Sequence[I+1] : I in 0..N-1],
  global_cardinality(Sequence,Pairs),

  solve([ff], Sequence).
```

Fig. 3.7 Magic sequence

the global_cardinality is *after* the extra constraints, then N = 400 is much faster, taking about 0.8 s.

The reason for this difference is that when Picat's cp module parses the constraint, it directly tries to reduce domains as much as possible. To see this more clearly, here are the domains after each constraint in the version of the model where global_cardinality follows the redundant constraints, when N = 7:

```
% Sequence = new_list(N),
% Sequence :: 0..N-1,
[0..6, 0..6, 0..6, 0..6, 0..6, 0..6, 0..6]


% N #= sum(Sequence),
[0..6, 0..6, 0..6, 0..6, 0..6, 0..6, 0..6]

% Integers = [I : I in 0..N-1],
% scalar_product(Integers, Sequence, N),
[0..6, 0..6, 0..3, 0..2, 0..1, 0..1, 0..1]


% create list: [0-Sequence[1], 1-Sequence[2], ...]
Pairs = [$I-Sequence[I+1] : I in 0..N-1],
global_cardinality(Sequence,Pairs),
[0..6, 0..6, 0..3, 0..2, 0..1, 0..1, 0..1]
```

After the `sum` constraint, no domains are reduced. However, after the `scalar_product` constraint, all of the domains for the numbers 2 and up are drastically reduced. The domains of the participating variables in the `global_cardinality` constraint are reduced. In the variant of the model where `global_cardinality` precedes the redundant constraints, this reduction is not done.

3.8 Circuit—Knight's Tour Problem

The *knight's tour problem*[11] is a well-known CSP problem, studied by Leonhard Euler. The original version of the problem is to create a *tour* on an 8×8 chessboard that covers all of the squares by using the movement of a knight: two steps in one non-diagonal direction followed by one step in an orthogonal direction. The knight's tour must be *closed*, meaning that from the last position in the tour it must be possible for the knight to jump to the starting square. Furthermore, each square on the chessboard must be visited exactly once. The general version of the problem is to create a knight's tour on square $N \times N$ chessboards that have sizes other than 8×8.

One example of a knight's tour on a 6×6 board is shown in Fig. 3.8. The tour starts at the upper-left square $(1, 1)$, marked by 1 (indicating the first step), continues to square $(2, 3)$, marked 2 (indicating the second step), etc. The last square covered is $(3, 2)$, marked with 36. Note that the 36 square is a knight's jump from the starting position, completing the tour.

The CP model shown in Fig. 3.9 will work with even-sized square boards, where $N \times N$ is an even number. However, for odd-sized boards, this model will not work. Note that, for $N < 6$, there are no solutions for even-sized square boards.

Problems that involve tours, which are also called *circuits*, can usually be modeled with the global constraint `circuit`.[12] The `circuit` constraint requires a list of decision variables with the domain `1..List.length`, and ensures that the list contains a circuit, where each value in the list represents the index of the next position to visit. Recall that, in a circuit, each position is visited exactly once, so each value in the list is different.

```
 1 32 29  6  3 18
30  7  2 19 28  5
33 36 31  4 17 20
 8 23 34 15 12 27
35 14 25 10 21 16
24  9 22 13 26 11
```

Fig. 3.8 Knight's tour on a 6×6 board

---

[11] https://en.wikipedia.org/wiki/Knight's_tour.
[12] http://sofdem.github.io/gccat/gccat/Ccircuit.html.

```
import cp.
main =>
  N = 6,
  knight(N,X),
  println(x=X),
  println("X:"),
  print_matrix(X),
  extract_tour(X,Tour),
  println("Tour:"),
  print_matrix(Tour).

% Knight's tour for even N*N.
knight(N, X) =>
  X = new_array(N,N),
  X :: 1..N*N,
  XVars = X.vars(),
  % restrict the domains of each square
  foreach (I in 1..N, J in 1..N)
    D = [-1,-2,1,2],
    Dom = [(I+A-1)*N + J+B : A in D, B in D,
           abs(A) + abs(B) == 3,
           member(I+A,1..N), member(J+B,1..N)],
    Dom.length > 0,
    X[I,J] :: Dom
  end,
  circuit(XVars),
  solve([ff,split],XVars).

extract_tour(X,Tour) =>
  N = X.length,
  Tour = new_array(N,N),
  K = 1,
  Tour[1,1] := K,
  Next = X[1,1],
  while (K < N*N)
    K := K + 1,
    I = 1+((Next-1) div N),
    J = 1+((Next-1) mod N),
    Tour[I,J] := K,
    Next := X[I,J]
  end.

print_matrix(M) =>
  N = M.length,
  V = (N*N).to_string().length,
  Format = "% " ++ (V+1).to_string() ++ "d",
  foreach(I in 1..N)
    foreach(J in 1..N)
      printf(Format,M[I,J])
    end,
    nl
  end,
  nl.
```

Fig. 3.9 Knight's tour model for even-sized square boards

An example of a circuit of length 6 is the list [2,3,4,5,6,1], indicating that, starting from position 1 (index 1), the next position visited is position 2, followed by positions 3, 4, 5, and 6. Finally, after position 6, position 1 is visited again in order to close the tour. The path, or tour, of the visited positions is 1,2,3,4,5,6.

A more interesting example is the circuit [3,5,2,1,6,4], where the visiting tour is 1,3,2,5,6,4. Note that the list that is created by the circuit constraint should not be confused with the visiting tour representation. Depending on the application, it might be necessary to convert the list given by the circuit constraint to show a more natural representation of the tour. The model in Fig. 3.9 is an example where such a conversion is needed.

The knight's tour model uses the matrix X as the main representation of the $N \times N$ board. However, the circuit constraint only works on one-dimensional lists, not on matrices, so the X matrix is converted to a list, XVars, via the vars/1 function.

The list XVars that is created by the circuit constraint represents the tour shown above:

```
XVars = [9,13,7,8,16,10,15,19,5,18,3,4,21,1,2,12,6,29,32,
         31,25,30,34,11,14,22,35,36,33,17,27,28,20,26,24,23]
```

The corresponding matrix representation, the matrix X, is shown below:

```
 9 13  7  8 16 10
15 19  5 18  3  4
21  1  2 12  6 29
32 31 25 30 34 11
14 22 35 36 33 17
27 28 20 26 24 23
```

The circuit shown in the X matrix and in XVars is interpreted in the same way as the lists shown above. The circuit always starts in position X[1,1] (or XVars[1]) which, in this case, contains the value of the second location that the knight visits: 9, which means the ninth position in the matrix, i.e. the (2, 3) square. The value of the ninth element in the list is 5 which means that the third location visited is in position 5, the (1, 5) square, etc.

This representation of the tour is much harder to follow than the representation shown in Fig. 3.8, so the model adds a helper predicate, extract_tour, to extract the visiting tour from the X matrix. In the while loop, each tour position, K that is stored in the X matrix (the value in the Next variable), is converted to the two-dimensional coordinates I and J, which are then inserted in the Tour matrix with the value of the corresponding K value, i.e. Tour[I,J] := K. The while loop ends when all $N \times N$ values have been converted and inserted into the Tour matrix.

The circuit constraint is the only constraint that is used in the model, but very important work is done in the foreach loop that precedes the circuit constraint. The foreach loop reduces the domain of each square so that it only contains the positions to which the knight can jump from that square. This domain reduction speeds up the model considerably. Note that the `circuit` constraint is placed **after** the domain-reducing loop in order to work with much smaller domains, rather than using all the possible values of `1..N`.

The predicate `print_matrix/1` is used for formatted printing of the matrices. Normally, printing the X matrix is not so interesting in a final model, and is included here for educational purposes. However, when **debugging** the model, it is important to print the X matrix.

The `circuit` constraint is used again in Chap. 7, where a CP model of the Traveling Salesman Problem is described and compared with other approaches.

Benchmarking: This model quickly solves most of the even-sized $N \times N$ square boards for $6 \leq N \leq 36$, and, with the exception of $N = 32$, all instances are solved well below 1 s; $N = 32$ is solved in about 35 s. For sizes larger than that, the model with the `[ff,split]` labeling has problems: for $N = 38$, the model does not solve the problem in 8 h. However, it does solve some larger instances in under 10 min, such as solving $N = 138$ in 94 s.

Using the `[ffc]` labeling instead, the $N = 38$ problem is solved in 0.5 s. However, it has the same behavior as the original labeling: while it solves some large problems quickly, such as solving $N = 72$ in 7 s, it has problems with smaller instances, such as $N = 24$, which is not solved within 10 min. Table 3.3 shows various sizes and the corresponding solve times for the two labelings `[ff,split]` and `[ffc]` with a 10-min timeout.

It is not easy to draw a conclusion based on these values, except that it is almost always worthwhile to experiment with different labelings for large problem instances. For quite a few instances one labeling times out, while the other solves it quickly. Another observation is that either a problem is solved in less than 1.5 min, or the problem is not solved before the 10-min timeout. With the exception of $N = 32$, the solving times increase with the problem size.

Table 3.3 Times for the knight's tour problem, even-sized $N \times N$ board, different labelings

N	ff,split (s)	ffc (s)
6	0.000	0.000
8	0.000	0.001
10	0.003	0.004
32	37.270	34.494
38	0.518	to
40	0.641	0.611
42	to	0.768
110	38.604	38.401
114	to	33.271
138	to	93.885

3.9 The `regular` Constraint—Global Contiguity and Nurse Rostering

The `regular` constraint is useful for certain kinds of scheduling problems where a schedule is a sequence of activities that is accepted by an automaton. A DFA (*deterministic finite automaton*) is given by a transition function, a start state, and a set of final states.

The hard part of using the `regular` constraint is to encode and set up the transition matrix. Later in this section, an example of a nurse rostering problem is demonstrated, but for explaining the `regular` constraint, a simpler problem is described first: a decomposition of the global constraint `global_contiguity`.

3.9.1 global_contiguity

In some models that use 0/1 variables, there is sometimes a need to ensure that, if there are any 1's, then they must appear in exactly one contiguous segment, for example [0,0,1,1,1,1,0,0,0] or [1,1,1,1,0,0,0,0,0]. The global constraint `global_contiguity`[13] ensures this property in a list of decision variables with the domains 0..1. This constraint is not a Picat built-in, so this section will define a decomposition of the constraint.

The requirement can be translated to the regular expression "0*1*0*", meaning that first there is a segment of zero or more 0's, then a segment of zero or more 1's, and finally a segment of zero or more 0's. The DFA for this is shown in Fig. 3.10.

State $S1$ is the start state, and $S3$ is the final state; final states are indicated by a double circle. From state $S1$, a 0 stays in $S1$, while a 1 transitions to state $S2$, etc.

The transition matrix for the DFA is shown in Table 3.4, where the rows represent the states, and the columns represent the input. The table encodes the same information as the graph, e.g., if in state $S1$ (row 1) and the input is "0", then stay in state $S1$; if the input is "1", then go to state $S2$ (row 2), etc.

In this representation, the state 0 represents a forbidden state, which means that it is not possible to be in state $S3$ with an input of "1".

This transition table can then be used to define the constraint `global_contiguity` with the following call to `regular`:

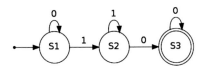

Fig. 3.10 DFA for "0*1*0*"

---

[13]http://sofdem.github.io/gccat/gccat/Cglobal_contiguity.html.

Table 3.4 Transition table
for global contiguity

State/Input	0	1
S1	1	2
S2	3	2
S3	3	0

```
regular(RegInput,NStates,InputMax,Transition,
        InitialState,FinalStates)
```

where the parameters are:

- `RegInput`: the list of decision variables with an appropriate domain (see the note below)
- `NStates`: the number of states, here 3
- `InputMax`: the number of inputs, here 2
- `Transition`: the transition matrix
- `InitialState`: the initial state, here state 1
- `FinalStates`: a list of final states. Here state 3 is the only final state.

Note: The `global_contiguity` constraint translates the values 0 and 1 of the input list to 1 and 2, since Picat's version of `regular` only works with positive integers in the `RegInput` list, meaning that 0 is not allowed.

The model's `main` predicate creates a list of six decision variables that have the domain of 0..1, calls the `global_contiguity` constraint, and then prints all possible solutions (Fig. 3.11).

3.9.2 Nurse Rostering

Now that the `regular` constraint has been explained, it is time to use the constraint for a scheduling problem.[14]

The *nurse rostering* scheduling problem has the following constraints:

- Each nurse must have one day off during every 4 days, and a nurse cannot work 3 nights in a row.
- There must be a minimum of `MinDayShift` nurses in each day shift, and a minimum of `MinNightShift` nurses in each night shift.

Figure 3.13 shows a model for the nurse rostering problem. The first constraint, the requirements for each nurse, uses the `regular` constraint as presented in Sect. 3.9.1, with the DFA in Fig. 3.12.

The three inputs are: o for a day off, d for a day shift, and n for a night shift. All of the states are final states, meaning that the schedule for a nurse can end in any state (any shift).

---

[14]This example was inspired by the nurse rostering example in the MiniZinc Tutorial.

```
import cp.

main =>
  N = 6,
  X = new_list(N),
  X :: 0..1,
  global_contiguity(X),
  solve(X),
  println(X),
  fail.

global_contiguity(X) =>
  N = X.length,

  % This uses the regular expression "0*1*0*" to
  % require that all 1's (if any) in an array
  % appear contiguously.
  Transition = [
                 [1,2], % state 1: 0*
                 [3,2], % state 2: 1*
                 [3,0]  % state 3: 0*
                 ],
  NStates = 3,
  InputMax = 2,
  InitialState = 1,
  FinalStates = [1,2,3],

  RegInput = new_list(N),
  RegInput :: 1..InputMax,   % 1..2

  % Translate X's 0..1 to RegInput's 1..2
  foreach (I in 1..N)
     RegInput[I] #= X[I]+1
  end,

  regular(RegInput,NStates,InputMax,
          Transition,InitialState,FinalStates).
```

Fig. 3.11 Global contiguity

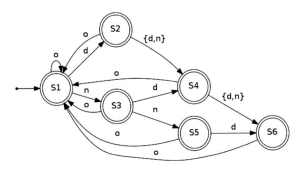

Fig. 3.12 DFA for nurse requirements

```
import sat.
main =>
   problem(1,NumNurses,NumDays,MinDayShift,MinNightShift),
   nurse_rostering(NumNurses,NumDays,
                   MinDayShift,MinNightShift,X,_Stat),
   foreach(Nurse in X) println(Nurse) end.

nurse_rostering(NumNurses,NumDays,
                MinDayShift,MinNightShift,X,Stat) =>
   % The DFA for the regular constraint.
   Transition = [
       % d n o
       [2,3,1], % state 1
       [4,4,1], % state 2
       [4,5,1], % state 3
       [6,6,1], % state 4
       [6,0,1], % state 5
       [0,0,1]  % state 6
   ],
   NStates      = Transition.length, % number of states
   InputMax     = 3,                 % 3 states
   InitialState = 1,                 % start at state 1
   FinalStates  = 1..6,              % all states are final
   DayShift   = 1,
   NightShift = 2,
   OffShift   = 3,
   % decision variables
   X = new_array(NumNurses,NumDays),
   X :: DayShift..OffShift,
   % summary of the shifts: [day,night,off]
   Stat = new_array(NumDays,3),
   Stat :: 0..NumNurses,
   % constraints
   foreach (I in 1..NumNurses)
      regular([X[I,J] : J in 1..NumDays],
              NStates,
              InputMax,
              Transition,
              InitialState,
              FinalStates)
   end,
   % statistics for each day
   foreach (Day in 1..NumDays)
      foreach (Type in 1..3)
         Stat[Day,Type] #=
            sum([X[Nurse,Day] #= Type : Nurse in 1..NumNurses])
      end,
      sum([Stat[Day,Type] : Type in 1..3]) #= NumNurses,
      % For each day the must be at least 3 nurses with
      % day shift, and 2 nurses with night shift
      Stat[Day,DayShift] #>= MinDayShift,
      Stat[Day,NightShift] #>= MinNightShift
   end,
   Vars = X.vars() ++ Stat.vars(),
   solve(Vars).

problem(1,NumNurses,NumDays,MinDayShift,MinNightShift) =>
   NumNurses     = 7,
   NumDays       = 14,
   MinDayShift   = 3, % minimum number in day shift
   MinNightShift = 2. % minimum number in night shift
```

Fig. 3.13 Nurse rostering, using the `regular` constraint

The DFA is then translated to the transition matrix that is shown in the model. The model has two matrices of decision values. X is the schedule for each nurse (the rows) for all the days in the schedule (the columns), where 1 represents a day shift, 2 represents a night shift, and 3 represents a day off. The Stat matrix collects the statistics of the days to ensure the requirements of minimum day and night shifts.

Modeling tip: When dealing with this type of DFA-based schedule, it is recommended to first use pen and paper to build the DFA as a graph, and then to write the transition matrix.

3.9.3 Alternative Approach for Valid Schedules: Table Constraint

An alternative approach for handling valid schedules is to use the table_in constraint (also known as *table*, *extensional constraint*, or *in_relation*[15]). The basic idea is to enumerate all the valid schedules, and then to ensure that a nurse's schedule is among these valid schedules (Fig. 3.14).

The table_in constraint has two parameters: Schedule, a tuple of days (as an array), and Valid, a list of all valid schedules.

An example is the following model, which has three valid 7-day schedules:

- [day,day,day,day,day,off,off]
- [night,off,night,off,day,day,off]
- [night,night,off,off,day,day,off]

including their rotational variants. In order to be used with table_in, these must be converted to a tuple of integers. The constraints regarding the minimum number of day and night shifts per day are the same as the regular variant that was described in Sect. 3.9.2.

For this specific scheduling example, the regular approach is probably better to use than the table_in approach, since the regular approach is more flexible. If the schedule should be changed to 14 days instead of 7, then the table_in variant must redefine the valid schedules, whereas this is not needed in the regular variant. Also, for the table_in model, all the rotational variants must be handled; this is not needed when using a DFA and regular.

Note: Picat also supports table_notin, which forbids certain combinations instead of allowing them.

---

[15] http://sofdem.github.io/gccat/gccat/Cin_relation.html.

```
import sat.

main =>
  Day   = 1, % day shift
  Night = 2, % night shift
  Off   = 3, % off
  % valid 7 day schedules
  Valid1 = [
             [Day,Day,Day,Day,Day,Off,Off],
             [Night,Off,Night,Off,Day,Day,Off],
             [Night,Night,Off,Off,Day,Day,Off]
           ],
  % create all rotational variants
  Valid = [],
  foreach (V in Valid1, R in 0..V.length-1)
    Rot = rotate_left(V,R).to_array(),
    Valid := Valid ++ [Rot]
  end,
  NumNurses = 14,
  NumDays = Valid[1].length,

  X = new_array(NumNurses,NumDays),
  X :: Day..Off,
  % ensure a valid scheme for each nurse
  foreach (Nurse in 1..NumNurses)
    table_in([X[Nurse,D] : D in 1..NumDays].to_array(),Valid)
  end,
  foreach (D in 1..NumDays)
    sum([X[Nurse,D] #= Day : Nurse in 1..NumNurses]) #>= 3,
    sum([X[Nurse,D] #= Night : Nurse in 1..NumNurses]) #>= 2
  end,
  Vars = X.vars(),
  solve(Vars),
  foreach(Nurse in X) println(Nurse) end.

rotate_left(L) = rotate_left(L,1).
rotate_left(L,N) = slice(L,N+1,L.length) ++ slice(L,1,N).
```

Fig. 3.14 Nurse rostering, using the `table_in` constraint

3.10 Debugging Constraint Models

Debugging a constraint model requires a different approach than debugging a traditional non-constraint program. The reason is that in a constraint model, everything "happens at the same time" when `solve` has been started. This means that the traditional "print debugging" does not work in the same way.

Some of the debugging methods have been mentioned earlier, such as printing the decision variables to see the domains in the read phase, or determining whether a certain point in the code is reached by using `println` statements.

The usual reason for debugging is that there is no solution, or that the given solution is not correct. Another reason may be that the model is too slow.

- Test early and often. For a beginner, it can be good practice to test the model after each new constraint is added.
- Check the domains.
- First test with a small instance that has a known output, in order to ensure that the model is correct. Testing is much simpler with small instances, since it will be faster to solve and easier to check the solutions.
- If a model doesn't work:

  – Remove one constraint after another and test again.
  – Check the domains again.

- In order to ensure correctness, count the number of solutions (using `find_all/2` or `solve_all`). For example, the 8-queens problem should give 92 solutions without symmetry breaking. A Sudoku problem should have exactly one solution: try to find 2 solutions by including `fail` at the end of the model.
- Use the `report` option to print intermediate values in optimization problems.
- If the (`cp`) model is too slow: Test different search strategies to speed up the search.

3.11 Bibliographical Note

The Global Constraint Catalog [6] is a large collection of the best-known global constraints, including descriptions of the constraints, references, and examples.

CSPLib [29] is a collection of many standard constraint satisfaction problems. This chapter quotes CSPLib problem 19, the magic squares and sequences problem [60], and problem 24, the Langford number problem [61]. For Picat implementations of some of the CSPLib problems, see http://www.csplib.org/Languages/Picat/models/.

MiniZinc is another language that can be used for constraint problems. The MiniZinc tutorial [38] includes a number of constraint modeling examples.

The implementation of each global constraint can require multiple algorithms. A discussion of the `regular` constraint and its implementation can be found in [43].

3.12 Exercises

1. Modify the program for Langford's number problem to solve the generalized variation: Given n sets of positive integers $1..k$, arrange them in a sequence such that, following the first occurrence of an integer i, each subsequent occurrence of i must appear exactly $i + 1$ indices after the prior occurrence.

2. Symmetry breaking: implement the `lex2(Matrix)`[16] constraint which ensures that all rows and columns in a matrix are lexicographically ordered. Hint: Use `lex_lt/2` or `lex_le/2`.

3. Determine why the Who Killed Agatha problem has eight different solutions.

4. The knight's tour model in Sect. 3.8 only works with even-sized square chessboards. Modify the model to support non-square $M \times N$ chessboards, where M and N are both even.

5. The knight's tour models in Sect. 3.8 and in the previous exercise only work with even-sized chessboards. Modify the model from Sect. 3.8 to support odd-sized chessboards, i.e. where $M \times N$ is odd.
   Hint 1: The global constraint `subcircuit` can be used.
   Hint 2: Most odd square sizes can be solved by excluding the middle square from the circuit and using the `circuit` constraint.

6. In the nurse rostering program, the day and night shifts for each nurse are not distributed evenly. Add to the model a way to minimize the difference between the numbers of day and night shifts. Hint: One way to do this is to minimize the differences of the "nurse points", where a nurse point is, for example, 1 for a day shift, and 2 for a night shift.

7. Test both the `cp` and `sat` modules for larger problem instances of the nurse rostering problem, such as more nurses and/or more days in the schedule. Also, change the minimum number of day/night shifts.

8. Implement the *3 jugs problem* using `regular`. The problem can be stated as:

   > There are 3 water jugs. The first jug can hold 3 liters of water, the second jug can hold 5 liters, and the third jug is an 8-liter container that is full of water. At the start, the first and second jugs are empty. The goal is to get exactly 4 liters of water in one of the containers.

9. Implement a *Nonogram* solver using `regular`. For a description of the problem, see https://en.wikipedia.org/wiki/Nonogram.

10. For each of the following global constraints, search the Global Constraint Catalog[17] (GCC) for its meaning. Determine whether Picat implements this constraint. If Picat implements the constraint, test it with the GCC's example. Otherwise, implement the constraint in Picat, and test it with the GCC's example.

    - `diffn/1`
    - `increasing/1`
    - `decreasing/1`
    - `nvalues/3`
    - `sliding_sum/4`

11. Implement the following problems from CSPLib[18]:

---

[16]http://sofdem.github.io/gccat/gccat/Clex2.html.
[17]http://sofdem.github.io/gccat/.
[18]http://www.csplib.org/Problems/.

- prob001: Car Sequencing
- prob006: Golomb rulers
- prob007: All-Interval Series
- prob016: Traffic Lights
- prob023: Magic Hexagon
- prob057: Killer Sudoku

Do not peek if there already is a Picat model.

Chapter 4
Dynamic Programming with Tabling

Abstract *Tabling* is a kind of memoization technique that caches the results of certain calculations in memory and reuses them in subsequent calculations through a quick table lookup. Tabling makes declarative dynamic programming possible. Users only need to describe how to break a problem into subproblems; they do not need to implement the data structures and algorithms for storing and looking up subproblems and their answers. This chapter describes the syntax and semantics of tabling in Picat, and demonstrates several dynamic programming examples.

4.1 Introduction

Dynamic programming (DP) is a recursive method for solving complex decision-making problems. DP employs two key ideas. First, it breaks a problem down into simpler subproblems, and constructs a solution for the main problem by combining the solutions of the subproblems. Second, it stores the solutions of all of the solved subproblems, so that the solution for a subproblem is only computed when the sub-problem occurs for the first time. This allows the stored solution to be reused for all of the subsequent occurrences of the same subproblem.

Picat provides a memoization technique, called *tabling*, for dynamic programming solutions. Tabling guarantees the termination of certain recursive programs. For example, consider the following non-tabled program:

```
reach(X,Y) ?=> edge(X,Y).
reach(X,Y) => reach(X,Z), edge(Z,Y).

% sample data
index (+,-)
edge(a,b).  edge(b,c).  edge(c,a).
```

where the predicate `edge` defines a relation, and the predicate `reach` defines the transitive closure of the relation. For a query, such as `reach(a,X)`, the program never terminates due to the existence of left-recursion in the second rule. Even if the rule is converted to right-recursion, the query still might not terminate if the graph contains cycles.

Tabling also removes redundancy in the execution of recursive programs. Section 1.4.4 showed that the tabled version of the Fibonacci function can be computed in linear time, while the non-tabled version takes exponential time to compute. As another example, consider the following problem: *Starting in the top left corner of an N × N grid, one can either go rightward or downward. How many routes are there through the grid to the bottom right corner?* The following gives a non-tabled function in Picat for the problem:

```
route(N,N,_Col) = 1.
route(N,_Row,N) = 1.
route(N,Row,Col) = route(N,Row+1,Col)+
                   route(N,Row,Col+1).
```

The function call `route(N,1,1)` takes exponential time in N, because the same function calls are repeatedly spawned during the execution, and are repeatedly resolved each time that they are spawned. If this function were tabled, then the call `route(N,1,1)` would take polynomial time, instead of exponential time.

4.2 Tabling in Picat

The idea of tabling is to memorize calls and their answers, and to use the stored answers to resolve a call if the same call has occurred before. In Picat, in order to table all of the calls and answers of a predicate or function, users just need to add the keyword `table` before the first rule. For example, the tabled versions of the `reach` and `route` programs are as follows:

```
table
reach(X,Y) ?=> edge(X,Y).
reach(X,Y) => reach(X,Z), edge(Z,Y).

table
route(N,N,_Col) = 1.
route(N,_Row,N) = 1.
route(N,Row,Col) = route(N,Row+1,Col)+
                   route(N,Row,Col+1).

% sample data
index (+,-)
edge(a,b).   edge(b,c).   edge(c,a).
```

With tabling, all queries to the `reach` predicate are guaranteed to terminate, and the function call `route(N,1,1)` takes only N^2 time.

For some problems, such as planning problems, it is infeasible to table every answer, because there may be a huge, or even an infinite, number of answers. For some other problems, such as those that require the computation of aggregates, it is a waste to table non-contributing answers. Picat allows users to provide *table modes* in order to instruct the system about which answers to table. For a tabled predicate, users can give a *table mode declaration* in the form $(M_1, M_2, \ldots, M_n)$, where each M_i is one of the following: a plus-sign (+) indicating input, a minus-sign (−) indicating output, min, indicating that the corresponding argument should be minimized, or max, indicating that the corresponding argument should be maximized. The last mode, M_n, can be nt, indicating that the corresponding argument will not be tabled.

Input arguments are normally ground. Output arguments, including min and max arguments, are assumed to be variables. An argument with the mode min or max is called an *objective* argument. Only one argument can be an objective to be optimized. As an objective argument can be a compound value, this limit is not essential, and users can still specify multiple objective variables to be optimized. When a table mode declaration is provided, Picat only tables one answer for each combination of input arguments. If there is an objective argument, then Picat only tables one optimal answer for each combination of input arguments.

A recursive predicate can be understood as a table of ground facts, called a *minimal model*,[1] that can be generated by the rules in the predicate. The minimal model may contain an infinite number of facts. For a tabled predicate, the same query may return different answers when under the control of different modes. Consider the following predicate:

```
table(+,max)
index (-,-)
p(a,2).
p(a,1).
p(a,3).
p(b,3).
p(b,4).
```

which is the same as:

```
table(+,max)
p(A1,A2)  ?=> A1=a,  A2=2.
p(A1,A2)  ?=> A1=a,  A2=1.
p(A1,A2)  ?=> A1=a,  A2=3.
p(A1,A2)  ?=> A1=b,  A2=3.
p(A1,A2)   => A1=b,  A2=4.
```

---

[1] A minimal model is the semantics given to programs that do not contain negation. There are several proposed definitions for the semantics of logic programs that contain negation. Since none of the examples presented in this book contains negation, the semantics of negation is not discussed in this book.

The query p(a,Y) returns the answer Y=3. If the table mode is changed to
(+,min), then the answer Y=1 is returned for the query. If the table mode is
replaced by (+,-), then the answer Y=2 is returned, since p(a,2) is the first fact
of p/2. An input argument with the mode (+) is normally ground; however, if it
is not ground, Picat returns an answer based on a different set of facts that can be
generated for the call. Using the above program as an example, the query p(X,Y)
returns the answer X=b, Y=4, which has the maximum value for Y among all of
the facts.

The nt mode should be used with caution. Two types of data can be passed to a
tabled predicate as an nt argument: (1) global data that are the same to every call of
the predicate, and (2) data that are functionally dependent on the input arguments.
When an nt mode exists, an answer is returned for a call as if the nt argument had
not been passed to the call and to the facts in the minimal model of the predicate.

4.3 The Shortest Path Problem

Given a weighted directed graph, a source node, and a destination node, a *shortest
path* from the source to the destination is a path with the minimum sum of the weights
of the constituent edges. Assume that the predicate edge(X,Y,W) specifies an
edge, where W is the weight of the edge from node X to node Y. The predicate
sp(X,Y,Path,W), defined below, states that Path is a path from X to Y with the
minimum total weight W.

```
table(+,+,-,min)
sp(X,Y,Path,W) ?=>
    Path = [(X,Y)],
    edge(X,Y,W).
sp(X,Y,Path,W) =>
    Path = [(X,Z)|Path1],
    edge(X,Z,Wxz),
    sp(Z,Y,Path1,W1),
    W = Wxz+W1.
```

Note that whenever the predicate sp/4 is called, the first two arguments must always
be instantiated. For each pair of nodes, the system only stores one path that has the
minimum weight.

The following program finds a shortest path among those that have the minimum
weight for each pair of nodes:

```
table(+,+,-,min)
sp2(X,Y,Path,WL) ?=>
    Path = [(X,Y)],
    WL = (Wxy,1),
```

```
   edge(X,Y,Wxy).
  sp2(X,Y,Path,WL) =>
    Path = [(X,Z)|Path1],
    edge(X,Z,Wxz),
    sp2(Z,Y,Path1,WL1),
    WL1 = (Wzy,Len1),
    WL = (Wxz+Wzy,Len1+1).
```

The predicate sp2(X,Y,Path,WL) returns a minimized term WL = (W,Len), where W is the total weight of a path from X to Y, and Len is the length of the path. Note that the order of W and Len in the pair is important. If the term would be (Len,W), then the program would find a shortest path, breaking a tie by selecting one with the minimum weight.

4.4 The Knapsack Problem

Given a set of items, each of which has a volume and a value, and a sack with a specified capacity volume, the *Knapsack problem* is to find a subset of items that can be put into the sack such that two goals are fulfilled: (1) the total volume of the items does not exceed the sack's capacity, and (2) the total value of the items is as large as possible.
  The following is a program that can solve the problem:

```
table(+,+,-,max)
knapsack(_,C,Sack,Val), C =< 0 =>
  Sack = [], Val = 0.
knapsack([_|L],C,Sack,Val) ?=>                    % Subprob 1
  knapsack(L,C,Sack,Val).
knapsack([Item@(IVol,IVal)|L],C,Sack,Val),  % Subprob 2
  C >= IVol
=>
  Sack = [Item|Sack1],
  knapsack(L,C-IVol,Sack1,Val1),
  Val = Val1 + IVal.
```

The predicate knapsack(Items,C,Sack,Val) has two input arguments and two output arguments. As input, it takes a list of Items, and C, the capacity of the sack. As output, it returns through Sack a list of items to be included in the sack, and returns through Val the total value of the included items. Each item in Items is a pair (IVol,IVal), where IVol is the item's volume and IVal is its value. For each item, there are two subproblems: one is to leave out the item, while the other is to include the item. The two subproblems are generated by two separate rules. Note that the rule for the first subproblem must be nondeterministic. If it were deterministic, then the last rule would never be applied.

The following gives a different program for the problem:

```
table
knapsack([],_,Sack,Val) =>
  Sack = [], Val = 0.
knapsack(_,C,Sack,Val), C =< 0 =>
  Sack = [], Val = 0.
knapsack([Item@(IVol,IVal)|L],C,Sack,Val),
  C >= IVol
=>
  knapsack(L,C,Sack1,Val1),              % Subprob 1
  knapsack(L,C-IVol,Sack2,Val2),         % Subprob 2
  (Val1 > Val2+IVal ->
    Sack = Sack1, Val = Val1
  ;
    Sack = [Item|Sack2], Val = Val2+IVal
  ).
knapsack([_|L],C,Sack,Val) =>
  knapsack(L,C,Sack,Val).
```

Unlike in the previous program, the two subproblems for an item are generated as a conjunction of recursive calls. The program takes the responsibility to compare the solutions of the two subproblems and to return the better solution. Since the predicate only contains deterministic rules and can only return one answer, table modes are not necessary.

4.5 The *N*-Eggs Problem

Given N eggs and a building of H floors, the aim of the *N-eggs problem* is to determine the highest floor from which an egg will not break when dropped to the ground, while minimizing the number, the number of drops used to determine the highest floor. All eggs are identical in terms of the shell's robustness. If an egg is dropped and does not break, its robustness is not affected, and the egg can be dropped again. However, once an egg is broken, it can no longer be dropped. This problem was used in the 2012 Prolog Programming Contest.

The following gives a program to find the minimum number of drops in the worst-case scenario:

```
table(+,+,min)
egg(_,0,NTries) => NTries=0.
egg(_,1,NTries) => NTries=1.
```

```
egg(1,H,NTries) => NTries=H.
egg(N,H,NTries) =>
  between(1,H,L),                     % choose a floor
  egg(N-1,L-1,NTries1),               % the egg breaks
  egg(N,H-L,NTries2),                 % the egg survives
  NTries is max(NTries1,NTries2)+1.
```

The predicate egg(N,H,NTries) takes as input N, the number of eggs, and H, the number of floors of the building. As output, the predicate returns through NTries the number of drops. The first three rules describe the base cases: the minimum number of tries is 0 when H = 0, 1 when H = 1, and H when N = 1. The last rule, which covers the general case, divides the problem into subproblems. The call between(1,H,L) nondeterministically selects a floor number L between 1 and H. There are two possible outcomes when dropping an egg from floor L. If the egg breaks, then the L-1 floors that are below floor L need to be tried, and the number of remaining eggs becomes N-1. If the egg does not break, then the H-L floors that are above floor L need to be tried, and the number of eggs remains the same. Since the worst-case scenario is considered, the number of tries for L is the maximum number from the two subproblems. The program returns the minimum of the numbers for all of the different choices of L.

The program only returns the minimum number of drops for any given values of N and H. It does not give a plan for these drops. The predicate can be extended with another argument to return a plan. This extension is left as an exercise.

4.6 Edit Distance

Given a string *p*, called a *pattern*, and another string *t*, called a *text*, the goal of the *edit-distance problem* is to find the shortest sequence of edit operations that transforms *p* into *t*. There are two types of edit operations:

- **Insertion**: Insert a single character into *p*.
- **Deletion**: Delete a single character from *p*.

The following program encodes the recursive algorithm for the problem:

```
table(+,+,min)
edit([],[],D) => D=0.
edit([X|P],[X|T],D) =>      % match
  edit(P,T,D).
edit(P,[_|T],D) ?=>         % insert
  edit(P,T,D1),
  D=D1+1.
edit([_|P],T,D) =>          % delete
  edit(P,T,D1),
  D=D1+1.
```

The program matches the text against the pattern from left to right. If the current character in the pattern matches the current character in the text, then the character is skipped. If they don't match, then there are two choices: one is to insert the current character from the text into the pattern, while the other choice is to delete the pattern's current character.

Note the argument D, which stores the edit distance between strings p and t. Each insertion and deletion operation increases this distance by 1. Since the objective argument min minimizes D, the total number of insertion and deletion operations is minimized.

The program takes $|p| \times |t|$ time, where $| \, |$ indicates the length of a string. Without tabling, the program would take exponential time.

4.7 The Longest Increasing Subsequence Problem

The *longest increasing subsequence* (LIS) problem is: given a sequence of numbers, find the longest possible subsequence in which the subsequence's elements are sorted in increasing order. The subsequence is not required to be contiguous or unique. For example, for the list [9,20,8,33,21,50,41,60,70], one LIS is [9,20,33,50,60,70].

The following gives a program for the problem:

```
table(+,-,max)
lis([],Sub,Len) => Sub = [], Len = 0.
lis([X|L],Sub,Len) ?=>  % Subprob 1: Sub starts here
    append(_,L1,L),
    lis(L1,Sub1,Len1),
    (Sub1 == [] -> true; Sub1 = [Y|_], X < Y),
    Sub = [X|Sub1],
```

Table 4.1 Execution time of lis on random lists

N	CPU time
500	0.03
1000	0.12
1500	0.28
2000	0.50
2500	0.79
3000	1.12
3500	1.57
4000	2.03
4500	2.57
5000	3.18

```
    Len = Len1+1.
  lis([_|L],Sub,Len) =>    % Subprob 2: Sub starts later
    lis(L,Sub,Len).
```

The predicate lis(L,Sub,Len) takes a list, L, and uses Sub and Len to return an LIS of L together with its length. In order to get an LIS of a list [X|L], this program divides the problem into two subproblems: one subproblem, generated by the second rule, assumes that the LIS begins with X, and the other subproblem, generated by the third rule, assumes that the LIS does not begin with X. In order to find an LIS that begins with X, the second rule further divides the problem into finding an LIS of each of the suffixes of L. The built-in predicate append(_,L1,L) is used to nondeterministically retrieve a suffix L1 from L. Let Sub1 be an LIS of L1. For [X|Sub1] to be an LIS of [X|L], Sub1 must either be empty, or must have a head Y that is greater than X.

For a list of N elements, the above program takes $O(N^2)$ time. The following test program can be used to confirm the time complexity.

```
  main =>
    Ns = 500..500..5000,   % from 500 to 5000 step 500
    member(N,Ns),
    initialize_table,
    L = [random() : _ in 1..N],
    statistics(runtime,_),
    lis(L,_Sub,_Len),
    statistics(runtime,[_,Time]),
    writeln((N,Time)), fail.
```

The failure-driven loop iterates through a list of different sizes, and, for each size, it generates a list of random numbers and calls lis on the list. The built-in initialize_table initializes the table area so that results from prior calls are not reused for the current call. Table 4.1 shows the execution time of the lis program on random lists.

4.8 Tower of Hanoi

The classic 3-peg *Tower of Hanoi problem* is posed as follows: Given three pegs and n disks of differing sizes stacked in increasing size order on the first peg, with the smallest disk on top, the goal is to stack the n disks on the last peg, using the following rules:

1. Only one disk can be moved at a time.
2. Only the top disk on any peg can be moved.
3. Larger disks cannot be stacked above smaller disks.

This problem can be solved algorithmically with $2^n - 1$ moves.

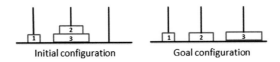

Initial configuration   Goal configuration

Fig. 4.1 Tower of Hanoi with configurations

This section considers a variant of the classic problem, where the initial and goal configurations are randomly generated. Figure 4.1 gives an instance of the problem.

The problem can be solved by recursively reducing it into subproblems. Let the disks be numbered $1, 2, \ldots, n$, from the smallest to the largest. For an initial configuration and a goal configuration, if disk n is on the same peg in both configurations, then the disk is removed. If $n = 0$ or $n = 1$, then the problem is solved or can be easily solved. Otherwise, the problem can be split into three subproblems. Let P_{n_i} be the peg on which disk n resides in the initial configuration, let P_{n_g} be the peg on which disk n resides in the goal configuration, and let P_o be the other peg that does not hold disk n in either the initial configuration or the goal configuration. The original problem can be solved recursively by the following three tasks: (1) move disks $1, 2, \ldots,$ and $n - 1$ to peg P_o; (2) move disk n from peg P_{n_i} to peg P_{n_g}; (3) move disks $1, 2, \ldots,$ and $n - 1$ from the configuration generated by the first task to the goal configuration.

Figure 4.2 gives a program for the 3-peg Tower of Hanoi problem. The predicate `hanoi(N, CState, GState, Plan, Len)` finds a plan that transforms the current state `CState` into the goal state `GState`. The mode declaration `(+, +, +, -, min)` for the predicate indicates that the number of disks and the states are input, the plan to be found is output, and the length of the plan should be minimized.

A state is represented as a triplet `{Peg1, Peg2, Peg3}`, where argument Peg_i stores a descendingly-sorted list of disks on peg i. For example, the following `main` predicate can be used to solve the instance shown in Fig. 4.1:

```
main =>
    hanoi(3,{[1],[3,2],[]},{[1],[2],[3]},Plan,Len),
    writeln(plan=Plan),
    writeln(len=Len).
```

The first rule in Fig. 4.2 handles the base case in which the number of disks is 0. The next three rules are for the case in which disk N is already on its final peg and can be removed from the configurations. The fifth rule handles another base case in which the number of disks is 1.

The last rule in Fig. 4.2 deals with the general case. The call

```
Pni = disk_on_peg(N,CState)
```

```
table (+,+,+,-,min)
hanoi(0,_CState,_GState,Plan,Len) => Plan=[],Len=0.
hanoi(N,{[N|Pi1],Pi2,Pi3},{[N|Pg1],Pg2,Pg3},Plan,Len) =>
      hanoi(N-1,{Pi1,Pi2,Pi3},{Pg1,Pg2,Pg3},Plan,Len).
hanoi(N,{Pi1,[N|Pi2],Pi3},{Pg1,[N|Pg2],Pg3},Plan,Len) =>
      hanoi(N-1,{Pi1,Pi2,Pi3},{Pg1,Pg2,Pg3},Plan,Len).
hanoi(N,{Pi1,Pi2,[N|Pi3]},{Pg1,Pg2,[N|Pg3]},Plan,Len) =>
      hanoi(N-1,{Pi1,Pi2,Pi3},{Pg1,Pg2,Pg3},Plan,Len).
hanoi(1,CState,GState,Plan,Len) =>
      Plan = [$move(Peg1,Peg2)],Len=1,
      Peg1 = disk_on_peg(1,CState),
      Peg2 = disk_on_peg(1,GState).
hanoi(N,CState,GState,Plan,Len) =>
      Pni = disk_on_peg(N,CState),
      Png = disk_on_peg(N,GState),
      Po = other_peg(Pni,Png),
      CState1 = remove_btm_disk(CState,Pni),
      N1 = N-1,
      IState = {_,_,_},
      IState[Po] = [I : I in N1..-1..1],
      IState[Pni] = [],
      IState[Png] = [],
      hanoi(N1,CState1,IState,Plan1,Len1),
      GState1 = remove_btm_disk(GState,Png),
      hanoi(N1,IState,GState1,Plan2,Len2),
      Plan = Plan1 ++ [$move(Pni,Png)|Plan2],
      Len = Len1+Len2+1.

remove_btm_disk({[_|P1],P2,P3},1) = {P1,P2,P3}.
remove_btm_disk({P1,[_|P2],P3},2) = {P1,P2,P3}.
remove_btm_disk({P1,P2,[_|P3]},_) = {P1,P2,P3}.

disk_on_peg(N,{[N|_],_,_}) = 1.
disk_on_peg(N,{_,[N|_],_}) = 2.
disk_on_peg(_,_) = 3.

other_peg(1,2) = 3.
other_peg(1,3) = 2.
other_peg(2,1) = 3.
other_peg(2,3) = 1.
other_peg(3,1) = 2.
other_peg(3,2) = 1.
```

Fig. 4.2 A program for Tower of Hanoi with configurations

returns the number of the peg on which disk N resides in CState. The call

```
Png = disk_on_peg(N,GState)
```

returns the number of the peg on which disk N resides in GState. The call

```
Po = other_peg(Pni,Png)
```

returns the number of the peg that is different from `Pni` and `Png`. Let `CState1` be the resulting state when removing disk `N` from `CState`, and let `GState1` be the resulting state when removing disk `N` from `GState`. The rule splits the `N`-disk problem into three subproblems. The call

```
hanoi(N1,CState1,IState,Plan1,Len1)
```

finds a plan that transforms `CState1` into `IState`, where `IState` holds `N-1` disks on peg `Po`, and no disks on pegs `Pni` and `Png`. The call

```
hanoi(N1,IState,GState1,Plan2,Len2)
```

finds a plan that transforms `IState` into `GState1`. After `Plan1` and `Plan2` are found, the step `move(Pni,Png)` is inserted between these two plans in order to move disk `N` from peg `Pni` to peg `Png`.

In the worst case, a plan that has an exponential number of steps in `N` may be required, causing the program to take exponential time in `N`. Without tabling, the program would always take exponential time in the length of the generated plan. With tabling, the program only takes polynomial time in the length of the generated plan. The current program treats symmetrical problems as different ones. For example, swapping any two of the pegs in both configurations of Fig. 4.1 will result in a completely different problem. The program can be improved by removing this kind of symmetry.

4.9 The Twelve-Coin Problem

The *twelve-coin problem* involves twelve coins, one of which is known to be counterfeit, while all of the others are identical genuine coins. It is known that the counterfeit coin weighs differently than a genuine coin, but it is unknown whether the counterfeit coin is heavier or lighter. The difference between the counterfeit coin and a genuine coin is only perceivable by weighing them on an old-style two-pan balance scale. The goal of the problem is to isolate the counterfeit coin with only three weighings.

A coin can be in one of the following possible states: U (unknown), PH (possibly heavier), PL (possibly lighter), H (heavier), L (lighter), or G (genuine). In the beginning, the state of every coin is U. The transitions are depicted in the following diagram:

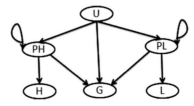

After a weighing, a U coin may become PH, PL, or G; a PH coin may stay unchanged or become H or G; and a PL coin may stay unchanged or become L or G. For example, consider a weighing in which three U coins are put on each side of the scale. The weighing gives one of the three possible outcomes. If both sides of the scale are equal, then all six of the U coins become G. If the left side is heavier than the right side, then the three U coins on the left become PH, and the three U coins on the right become PL. The third outcome, in which the right side is heavier than the left side, is similar to the second outcome.

A state is represented by the array {NU,NPH,NPL,NG,NW}, where NU is the number of unknown coins, NPH is the number of possibly heavier coins, NPL is the number of possibly lighter coins, NG is the number of genuine coins, and NW is the remaining number of weighings. For the twelve-coin problem, the initial state is represented by {12,0,0,0,3}. In this representation, the coins of the same type are not distinguishable, and the number of coins of each type is the only concern. This removal of symmetry significantly reduces the number of possible combinations of coins for weighing.

Figure 4.3 gives a program for the problem. The program uses the CP solver to generate different possible weighings for each state. For a state

```
{NU,NPH,NPL,NG,NW}
```

assume that the numbers of U, PH, PL, and G coins to be put on the left side of the scale are, respectively, NU1, NPH1, NPL1, and NG1, and that the numbers of coins to be put on the right side are, respectively, NU2, NPH2, NPL2, and NG2. Constraints are used in order to guarantee that: (1) the number of coins of each type to be put on the scale does not exceed the total number of the type, and (2) both sides of the scale have an equal number of coins, which must not be zero. The lex_le constraint ensures that the left side is lexicographically less than or equal to the right side. This constraint removes another type of symmetry.

After a weighing, the original problem is divided into three subproblems. The first subproblem deals with the outcome in which both sides are equal. Because all of the coins that were just weighed are known to be genuine coins, the number of remaining U coins is NU−NU1−NU2, the number of PH coins is NPH−NPH1−NPH2, the number of PL coins is NPL−NPL1−NPL2, and the number of G coins is the sum of the original number of G coins plus the number of coins that were just weighed.

The second subproblem results from the outcome that the left side is heavier than the right side. In this outcome, no coins will remain being U, because all of the U coins on the left become PH, all of the U coins on the right become PL, and all of the coins that were not weighed become G. Also, all of the PL coins on the left and all of the PH coins on the right become G.

The third subproblem, which deals with the outcome that the right side is heavier than the left side, is similar to the second subproblem.

The following test program can be used to find the minimum number of weighings, M, needed for a given number of coins N. Table 4.2 gives the results, which were obtained on PC notebook with 2.4 GHz Intel i5 and 8 GB RAM.

```
import cp.

main =>
  coin({12,0,0,0,3},Plan),
  writeln(Plan).

table(+,-)
coin({0,0,0,_,_},Plan) => Plan=nil.
coin({0,1,0,_,_},Plan) => Plan=nil.
coin({0,0,1,_,_},Plan) => Plan=nil.
coin(State@{NU,NPH,NPL,NG,NW},Plan),NW>0 =>
  Plan = $weigh(State,
                {NU1,NPH1,NPL1,NG1},
                {NU2,NPH2,NPL2,NG2},Plan1,Plan2,Plan3),
  [NU1,NU2]   :: 0..NU,
  [NPH1,NPH2] :: 0..NPH,
  [NPL1,NPL2] :: 0..NPL,
  [NG1,NG2]   :: 0..NG,
  NU1+NU2 #=< NU,
  NPH1+NPH2 #=< NPH,
  NPL1+NPL2 #=< NPL,
  NG1+NG2 #=< NG,
  LSide #= NU1+NPH1+NPL1+NG1,
  RSide #= NU2+NPH2+NPL2+NG2,
  lex_le([NU1,NPH1,NPL1,NG1],[NU2,NPH2,NPL2,NG2]),
  LSide #> 0,
  RSide #> 0,
  LSide #= RSide,        % equal number of coins on both sides
  NW1 = NW-1,
  solve([NU1,NU2,NPH1,NPH2,NPL1,NPL2,NG1,NG2]),
  O1 = {NU-NU1-NU2,      % Outcome1: both sides are equal
        NPH-NPH1-NPH2,
        NPL-NPL1-NPL2,
        NG+NU1+NPH1+NPL1+NU2+NPH2+NPL2,
        NW1},
  coin(O1,Plan1),
  O2 = {0,              % Outcome2: left side is heavier
        NPH1+NU1,
        NPL2+NU2,
        NG+(NPH-NPH1)+(NPL-NPL2)+(NU-NU1-NU2),
        NW1},
  coin(O2,Plan2),
  O3 = {0,              % Outcome3: left side is lighter
        NPH2+NU2,
        NPL1+NU1,
        NG+(NPH-NPH2)+(NPL-NPL1)+(NU-NU1-NU2),
        NW1},
  coin(O3,Plan3).
```

Fig. 4.3 A Picat program for the twelve-coin problem

N	M	CPU time
12	3	0.00
20	4	0.15
30	4	1.61
40	5	11.81
50	5	52.12
60	5	182.21
70	5	525.07

Table 4.2 Minimum number of weighings for N-coins

```
main =>
  member(N,[12,20,30,40,50,60,70]),
  statistics(runtime,_),
  once minW(N,M),
  statistics(runtime,[_,Time]),
  writeln((N,M,Time)), fail.

minW(N,M) =>
  initialize_table,
  between(1,N,M),
  coin({N,0,0,0,M},_).
```

4.10 Bibliographical Note

Dynamic programming is an important problem-solving method, which is covered by many textbooks on data structures and algorithms (e.g., [15]). Memoization was first used in functional programming to speed up the evaluation of recursive functions [41]. OLDT [56] was the first extension of SLD resolution [36] with memoization that aimed to cure infinite loops in SLD resolution. OLDT relies on suspension of calls to compute the fixed points. OLDT was formalized as SLG [12] and implemented in XSB [47]. Initial applications of tabled logic programming can be found in [64]. These applications attracted more systems to support tabling [48, 68].

Linear tabling [76], which relies on iterative computation of looping calls, was proposed as a simpler method than SLG for evaluating tabled logic programs. Mode-directed tabling [22] was designed for solving dynamic programming problems. The nt mode, which facilitates passing global or functionally dependent data to required computations, was proposed in [75]. In Picat, ground structured terms in tabled calls are hash-consed [74] so that common ground terms are only tabled once. This optimization greatly improves the scalability of dynamic programming solutions in Picat.

Mode-directed tabling, as implemented in B-Prolog [68], was found useful for some of the ASP competition problems [11], such as the Sokoban problem [71] and the 4-peg Tower of Hanoi problem [73]. These successes led to continued research about tailoring tabling to planning. Other dynamic programming problems, including some of the exercises in this chapter, are located at [17].

4.11 Exercises

1. Modify the N-eggs program from Sect. 4.5 such that, in addition to returning the minimum number of drops in the worst scenario, it also returns a plan that has the form drop(L, Plan1, Plan2), where L is the number of a floor from which an egg is dropped, Plan1 is a subplan for the case in which the egg breaks, and Plan2 is a subplan for the case in which the egg does not break.
2. Modify the LIS program from Sect. 4.7 to resolve ties as follows: If the given sequence has more than one possible LIS, then the program will return the LIS whose sum is largest.
3. Write a function that takes a list of real numbers $[E_1, \ldots, E_n]$ and returns the non-empty contiguous sublist $[E_i, \ldots, E_j]$ that has the largest possible sum.[2]
4. Modify the function from Exercise 3 to work with two-dimensional lists. Given a two-dimensional list of real numbers, with dimensions m and n, return the non-empty contiguous two-dimensional sublist that has the largest possible sum.
5. Write a program to solve the *box stacking problem* (see Footnote 2). Given a list of n rectangular 3-D boxes, each of which has a Height, a Width, and a Depth that are specified as real numbers, the goal is to make a stack of boxes that is as tall as possible. However, a box can only be put on top of another box if the dimensions of the 2-D base of the lower box are each strictly larger than those of the 2-D base of the higher box. Each box can be rotated so that any side functions as its base. There can be multiple boxes with the same dimensions.
6. Write a program to partition a set of n natural numbers into two non-empty subsets in a way that minimizes $|S_1 - S_2|$, where S_1 and S_2 are the sums of the elements in each of the two subsets (See Footnote 2).
7. Consider a row of n coins of given values $v_1 \ldots v_n$, where n is even. You are playing a game in which you alternate turns with an opponent. In each turn, a player selects either the first or last coin from the row, permanently removes it from the row, and receives the value of the coin. Write a program to determine the maximum amount of money that you can win if you move first.
8. Write a program to solve the *all-pairs shortest paths* problem. Given a weighted directed graph of n nodes that is defined by a series of predicates edge(X, Y, W), as described in Sect. 4.3, the program returns an $n \times n$ list, L, where each L[X, Y] contains the weight of the shortest path from node X to node Y.

---

[2] Taken with permission from http://people.cs.clemson.edu/~bcdean/dp_practice.

9. Rewrite the program shown in Fig. 4.2 for the 3-peg Tower of Hanoi problem to allow symmetrical problems to share solutions.

10. Consider the *4-peg Tower of Hanoi problem* with given initial and goal configurations. This problem can be solved by using dynamic programming, similarly to how the 3-peg version was solved in Sect. 4.8. For a problem with n disks, if disk n is on the same peg in both the current and goal configurations, then the disk can be removed from consideration. Otherwise, if $n > 1$, the *Frame-Stewart algorithm* can be used to divide the problem into the following subproblems:

   a. Move the *Mid* topmost disks to another, intermediate, peg (which is not the destination peg), using all 4 pegs.
   b. Move the $n - Mid$ remaining disks to the destination peg using the 3 remaining pegs.
   c. Move the *Mid* disks from their current peg to the destination peg, using all 4 pegs.

   Write a program to solve the 4-peg Tower of Hanoi problem.

Chapter 5
From Dynamic Programming to Planning

Abstract Planning can be treated as dynamic programming, for which tabling has been shown to be effective. Picat provides a module, named `planner`, which is based on tabling, but provides a level of abstraction that hides tabling from users. This chapter focuses on depth-unbounded search predicates in the `planner` module, and demonstrates several application examples.

5.1 Introduction

Planning means finding a sequence of activities required to achieve a desired goal. Planning is indispensable for many tasks, such as logistics, robotics, and military operations. Many puzzles, such as the fifteen sliding puzzle and the Rubik's cube puzzle, are planning problems. The *classic planning problem* is the most primitive form of planning. Given an initial state, a description of the final states, and a set of possible actions, the classic planning problem is to find a plan that transforms the initial state to a final state.

The classic planning problem can be viewed as a path-finding problem over an implicitly-specified graph. The tabled program given in Chap. 4 for the shortest-path problem can be generalized for classic planning. The following shows the framework:

```
table (+,-,min)
path(S,Plan,Cost),final(S) =>
    Plan=[],Cost=0.
path(S,Plan,Cost) =>
    action(S,NextS,Action,ACost),
    path(NextS,Plan1,Cost1),
    Plan = [Action|Plan1],
    Cost = Cost1+ACost.
```

The call `path(S, Plan, Cost)` binds `Plan` to an optimal plan that can transform state `S` to a final state. The predicate `final(S)` succeeds if `S` is a final state, and the predicate `action/4` encodes the set of actions in the problem. For an initial state, the predicates `final/1` and `action/4` constitute a *state space*. Picat performs

© The Author(s) 2015
N.-F. Zhou et al., *Constraint Solving and Planning with Picat*,
SpringerBriefs in Intelligent Systems, DOI 10.1007/978-3-319-25883-6_5

depth-first tabled search in the state space, meaning that Picat tables every state that is encountered during the search in order to avoid exploring a state more than one time.

Picat provides a module, named `planner`, which provides a level of abstraction that hides tabling from users. For a planning problem, in order to find a plan or an optimal plan, users only need to define the predicates `final/1` and `action/4`, and call one of the `planner` module's search predicates, passing an initial state as a parameter. In addition to performing *depth-unbounded search*, as described in the framework above, the `planner` module also provides predicates for performing *depth-bounded search*, such as *iterative-deepening* and *branch-and-bound*. This chapter describes depth-unbounded search predicates in the `planner` module, together with their implementations and several application examples. Chapter 6 covers depth-bounded search and heuristic search for planning.

5.2 The `planner` Module: Depth-Unbounded Search

Picat's `planner` module provides predicates and functions for specifying and solving planning problems. In order to solve a planning problem, a program must import the `planner` module and provide the following two global predicates:

- `final(S)`: This predicate succeeds if S is a final state.
- `action(S, NextS, Action, ACost)`: This predicate encodes the *state transition diagram* of a planning problem. State S can be transformed to $NextS$ by performing *Action*. The cost of *Action* is *ACost*, which must be non-negative. If the plan's length is the only interest, then $ACost=1$. Note that the assignment operator `:=` cannot be used to update variable S to produce `NextS`.

The `final` and `action` predicates are called by the planner. The `action` predicate specifies the precondition, effect, and cost of each of the actions. This predicate is normally defined with nondeterministic pattern-matching rules. The planner tries actions in the order in which they are specified. When a non-backtrackable rule is applied to a call, the remaining rules will be discarded for the call.

The following describes the predicates in the `planner` module that perform depth-unbounded search:

- `best_plan_unbounded(S, Limit, Plan, PlanCost)`: This predicate, if it succeeds, binds *Plan* to a plan that can transform state S to a final state. *PlanCost* is the cost of *Plan*. *PlanCost* cannot exceed *Limit*, which is a given non-negative integer. The argument *PlanCost* is optional. If it is omitted, then the predicate does not return the plan's cost. The *Limit* can also be omitted. In this case, the predicate assumes the cost limit to be 268435455.
- `plan_unbounded(S, Limit, Plan, PlanCost)`: This predicate is the same as the `best_plan_unbounded` predicate, except that it terminates the search once it finds a plan whose cost does not exceed *Limit*.

5.3 The Implementation of Depth-Unbounded Search

Figure 5.1 shows Picat's implementation of the depth-unbounded search predicates. For a call to best_plan_unbounded, Picat explores a path until it encounters either a final state, a *dead-end* state in which no action can be applied, or a state that has occurred before. When Picat encounters a dead-end or a recurring state, it backtracks. Picat tables all of the encountered states so that a state that occurs in multiple paths in the search space is only expanded once during each round of

```
best_plan_unbounded(S,Limit,Plan,PlanCost) =>
  best_plan_unbounded_aux(S,Plan,PlanCost),
  PlanCost =< Limit.

table (+,-,min)
best_plan_unbounded_aux(S,Plan,PlanCost),final(S) =>
  Plan = [],
  PlanCost = 0.
best_plan_unbounded_aux(S,Plan,PlanCost) =>
  action(S,NextS,Action,ACost),
  best_plan_unbounded_aux(NextS,Plan1,PlanCost1),
  Plan = [Action|Plan1],
  PlanCost = PlanCost1+ACost.

%%%
plan_unbounded(S,Limit,Plan,PlanCost) =>
  IPlan = {Limit,[],0},
  catch(plan_unbounded_aux(S,Plan,PlanCost,IPlan),
        (Plan,PlanCost), true).

table (+,-,min,nt)
plan_unbounded_aux(S,Plan,PlanCost,_),
  final(S)
=>
  Plan = [],
  PlanCost = 0.
plan_unbounded_aux(S,Plan,PlanCost,{Limit,IPlan,IPlanCost}) =>
  action(S,NextS,Action,ACost),
  Inherited1 = {Limit-ACost,
                [Action|IPlan],
                IPlanCost+ACost},
  plan_unbounded_aux(NextS,Plan1,PlanCost1,Inherited1),
  Plan = [Action|Plan1],
  PlanCost = PlanCost1+ACost,
  (PlanCost =< Limit ->
      throw((IPlan.reverse()++Plan,IPlanCost+PlanCost))
  ;
      true
  ).
```

Fig. 5.1 The implementation of the depth-unbounded search predicates

evaluation of the call. Picat keeps replacing existing plans for the tabled states with better plans until a round of search fails to find any better plans. Note that the argument *Limit* is not used to limit the depth of the search. Instead, *Limit* is compared with *PlanCost* after an optimal plan has been found. During depth-unbounded search, once a state has failed, it will not be explored again, whether it occurs in the current round or in subsequent rounds.

In the implementation of `plan_unbounded`, the following array is passed from a state to the next state:

```
{Limit,IPlan,IPlanCost}
```

`Limit` is the maximum amount of resources that can be consumed by future actions. `IPlan` is a partial plan that is inherited from the parent. `IPlanCost` is the total cost of `IPlan`. Initially, `Limit` is the limit that is given by the user, `IPlan` is [], and `IPlanCost` is 0. The array is passed down as an nt argument. Therefore, it is not used in variant checking. Two calls to `plan_unbounded_aux` are variants if the states that they represent (i.e., the first arguments of the calls) are the same. Once a plan is found for the initial state whose cost does not exceed the limit, the program throws the plan to the initial call as an exception, terminating the search.

5.4 Missionaries and Cannibals

The *missionaries and cannibals problem* is a classic AI planning problem. Three missionaries and three cannibals come to the southern bank of a river and find a boat that holds up to two people. If the cannibals ever outnumber the missionaries on either bank, the missionaries will be eaten. Find a plan to move them to the other bank of the river.

Figure 5.2 gives a program for the problem. A state is represented as a list of the form [M, C, Bank]. Bank is the bank of the river where the boat is currently located. M and C are respectively the number of missionaries and the number of cannibals on the Bank of the river. Initially, all of the missionaries and cannibals, as well as the boat, are on the southern bank of the river.

The `final` predicate gives the condition for the final state. The current state is final if the boat and all six people are on the northern bank of the river.

The `action` predicate is defined with a single rule. For the current state [M, C, Bank], a number of missionaries, BM, is chosen from the range 0..M, and a number of cannibals, BC, is chosen from the range 0..C to cross the river. For the crossing to be valid, the total number of people on the boat must either be 1 or 2. Furthermore, if there is a missionary on either bank, the cannibals cannot outnumber the missionaries. After crossing, the boat comes to the opposite bank, and the program updates the numbers of missionaries and cannibals on each bank.

The program, using Picat version 1.3, finds an optimal plan of 11 steps for the problem in 0.04 s on an Intel-i5 PC.

```
import planner.

main =>
    best_plan_unbounded([3,3,south],Plan),
    foreach (Step in Plan)
        println(Step)
    end.

final([3,3,north]) => true.

action([M,C,Bank],NextS,Action,Cost) =>
    member(BM,0..M),
    member(BC,0..C),
    BM+BC > 0, BM+BC =< 2,
    OppBank = opposite(Bank),
    Action = $cross(BM,BC,OppBank),
    Cost = 1,
    NewM1 = M-BM,
    NewC1 = C-BC,
    NewM2 = 3-NewM1,
    NewC2 = 3-NewC1,
    if NewM1 !== 0 then    % missionaries are safe
        NewM1 >= NewC1
    end,
    if NewM2 !== 0 then
        NewM2 >= NewC2
    end,
    NextS = [NewM2,NewC2,OppBank].

opposite(south) = north.
opposite(north) = south.
```

Fig. 5.2 A program for the Missionaries and Cannibals problem

5.5 Klotski

Klotski is a sliding puzzle. There are one 2×2 piece, one 2×1 piece, four 1×2 pieces, and four 1×1 pieces. Initially, the pieces are placed on a 4×5 board, as shown in Fig. 5.3. After the pieces are placed, the board has two empty spaces. The goal of the game is to slide the 2×2 piece to the exit, so that the 2×2 piece can be slid out through the exit during the next move. No other pieces can be removed from the board. Pieces must be slid horizontally or vertically to empty spaces. The location of a piece on the board is represented as (R, C), where R is the number of the row and C is the number of the column of upper-left square of the piece. Therefore, the 2×2 piece is at the exit if its upper-left corner is in position (4,2).

Figure 5.4 gives a Picat program for the puzzle. A state is represented as a pair {Spaces, Pieces}, where Spaces is a sorted list of locations of the two empty spaces, and Pieces is a sorted list of pieces. Each piece is represented as a pair, where the first element is the size of the piece, and the second element is the location

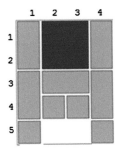

Fig. 5.3 Klotski

of the piece. In the `move_piece` predicate, this pair corresponds to the second and third arguments, `D@(W,H)` and `P@(R,C)`. Recall that a pattern that is preceded by @ denotes an *as-pattern*. The as-pattern `D@(W,H)` is the same as the pattern `(W, H)`, except that *D* is made to reference the term that matched the pattern.

The initial state, shown in Fig. 5.3, can be defined as follows:

```
initial(S) =>
    S = {[(5,2), (5,3)],          % empty spaces
         sort([{(2,2),(1,2)},      % one 2 x 2 piece
               {(2,1),(3,2)},      % one 2 x 1 piece
               {(1,2),(1,1)},      % four 1 x 2 pieces
               {(1,2),(3,1)},
               {(1,2),(1,4)},
               {(1,2),(3,4)},
               {(1,1),(5,1)},      % four 1 x 1 pieces
               {(1,1),(4,2)},
               {(1,1),(4,3)},
               {(1,1),(5,4)}])}.
```

A state is final if the 2 × 2 piece is at location `(4, 2)`. This condition is represented by `member((2,2), (4,2), Pieces)`. Note that the condition represents multiple final states, because although the 2 × 2 square is required to be moved to `(4,2)`, the other pieces can be at any locations.

The `action` predicate is defined by one rule. First, the rule nondeterministically selects a piece. Then, it calls `move_piece` to move the piece. A piece can be moved in one of four directions: up, down, left, and right. There must be space for the piece to move in the chosen direction. For example, consider the first `move_piece` rule, which moves a piece left from `(R,C)` to `(R,C1)`, where `C1 = C-1`. If the height of the piece is 1, meaning that `H == 1`, then `select` checks that one of the empty spaces is at `(R,C1)`. After the move, the new location of the piece is `(R,C1)`, and the empty space changes to location `(R,C1+W)`, where `W` is the width of the piece. Otherwise, if the height of the piece is 2, then the rule checks that the two empty spaces are vertically adjacent to the left edge of the piece, at locations `(R,C1)` and

```
import planner.

main =>
    initial(S0),
    best_plan_unbounded(S0,Plan),
    foreach (Step in Plan)
       println(Step)
    end,
    println(len=length(Plan)).

final({_,Pieces}), member({(2,2),(4,2)},Pieces) => true.

action({Spaces,Pieces},NextS,Action,Cost) =>
    Cost = 1,
    select({Dim,Pos},Pieces,Pieces1),
    move_piece(Spaces,Dim,Pos,NSpaces,NPos,Action),
    NextS = {NSpaces,insert_ordered(Pieces1,{Dim,NPos})}.

move_piece(Spaces,D@(W,H),P@(R,C),NSpaces,NP,Action) ?=>
    Action = $move(D,P,left),
    C1 = C-1, NP = (R,C1),
    if H==1 then
        select((R,C1),Spaces,SpacesR),
        NSpaces = insert_ordered(SpacesR,(R,C1+W))
    else
        Spaces = [(R,C1),(R+1,C1)],
        NSpaces = [(R,C1+W),(R+1,C1+W)]
    end.
move_piece(Spaces,D@(W,H),P@(R,C),NSpaces,NP,Action) ?=>
    Action = $move(D,P,right),
    C1 = C+1, NP = (R,C1),
    if H==1 then
        select((R,C+W),Spaces,SpacesR),
        NSpaces = insert_ordered(SpacesR,(R,C))
    else
        Spaces = [(R,C+W),(R+1,C+W)],
        NSpaces = [(R,C),(R+1,C)]
    end.
move_piece(Spaces,D@(W,H),P@(R,C),NSpaces,NP,Action) ?=>
    Action = $move(D,P,up),
    R1 = R-1, NP = (R1,C),
    if W==1 then
        select((R1,C),Spaces,SpacesR),
        NSpaces = insert_ordered(SpacesR,(R1+H,C))
    else
        Spaces = [(R1,C),(R1,C+1)],
        NSpaces = [(R1+H,C),(R1+H,C+1)]
    end.
move_piece(Spaces,D@(W,H),P@(R,C),NSpaces,NP,Action) =>
    Action = $move(D,P,down),
    R1 = R+1, NP = (R1,C),
    if W==1 then
        select((R+H,C),Spaces,SpacesR),
        NSpaces = insert_ordered(SpacesR,(R,C))
    else
        Spaces = [(R+H,C),(R+H,C+1)],
        NSpaces = [(R,C),(R,C+1)]
    end.
```

Fig. 5.4 A program for Klotski

(R+1,C1), respectively. After the move, the new locations of the two empty spaces are (R,C1+W) and (R+1,C1+W), respectively.

The program, using Picat version 1.3, finds an optimal plan of 116 moves in 15 s on an Intel-i5 PC. As sliding a piece to an empty space and then sliding it to the next empty space constitutes two moves in the encoding, the solution lists more steps than the known 81-step minimum-length solution for the puzzle.

5.6 Sokoban

Sokoban is a type of transport puzzle, in which the player, named sokoban (which means "warehouse-keeper" in Japanese), must push all the boxes on the floor to the designated storage locations. Only one box may be pushed at a time, and boxes cannot be pulled. This problem has been shown to be NP-hard, and has raised great interest because of its relation to robot planning.

In standard Sokoban, the player solves the puzzle by pushing boxes. However, boxes can get stuck after being pushed to certain locations, resulting in unsolvable configurations. For this reason, it is easier to solve the puzzle backward from the goal state to the initial state by pulling boxes instead of pushing them. Any plan for the reversed puzzle can be converted to a plan for the original puzzle. Although boxes can still get stuck, reverse Sokoban has fewer unsolvable configurations than standard Sokoban. Figure 5.5 gives a problem instance and its reverse.

The floor configuration of an instance is defined by the following static facts:

- size(NRows, NCols): The floor has NRows rows and NCols columns.
- wall(R,C): The square at (R,C) is a wall. Only interior walls need to be specified. The boundary is assumed to have walls. Neither boxes nor the sokoban can enter squares that are occupied by walls.
- goal(R,C): The square at (R,C) is a goal location.

A state is represented by a pair of the form {SoLoc, BoxLocs}, where SoLoc is the sokoban's current location, and BoxLocs is a sorted list of the boxes' locations. This representation automatically removes box symmetries. Since the repre-

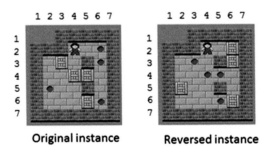

Original instance   Reversed instance

Fig. 5.5 A Sokoban instance and its reverse

sentation does not carry the identities of boxes, two configurations are represented as the same state if they are not distinguishable by the boxes' locations.

For an instance, the `main` predicate loads the static facts into the system, and calls the planner on an initial state. For the reversed Sokoban instance that is shown in Fig. 5.5, the `main` predicate is the following:

```
main =>
    Facts = $[size(7,7),
              wall(2,2), wall(2,3), wall(3,5),
              wall(5,5), goal(3,3), goal(4,4),
              goal(4,5), goal(6,5)],
    cl_facts(Facts,$[wall(+,+), goal(+,+)]),
    SoLoc = (2,4),
    BoxLocs = [(2,6),(3,6),(5,2),(6,6)],
    best_plan_unbounded({SoLoc,BoxLocs},Plan),
    foreach (Step in Plan)
       println(Step)
    end.
```

The built-in `cl_facts` loads the facts into the system, assuming that all of the arguments of calls to `wall` and `goal` are ground.

Figure 5.6 gives definitions of the `final` and `action` predicates. A state is final if every location in `BoxLocs` is a goal location.

```
final({_SoLoc,BoxLocs}) =>
    foreach((R,C) in BoxLocs)
       goal(R,C)
    end.
```

Note that the condition represents multiple final states, because although the stones must be pushed to the designated goal locations, the `sokoban` himself can be at any location.

The actions are defined by two rules. The first rule of `action/4` selects a box that is next to the `sokoban`, chooses a direction and an empty square in that direction, and pulls the box to the empty square. The `neib(SoLoc, BoxLoc, Dir)` predicate finds the `BoxLoc` that is next to `SoLoc` in the direction `Dir`. The `select(BoxLoc, BoxLocs, BoxLocs1)` predicate ensures that `BoxLoc` is an element in the list `BoxLocs`, and removes `BoxLoc` from `BoxLocs`, producing `BoxLocs1`. In order to pull the box at `BoxLoc` in `OppDir`, which is the opposite direction of `Dir`, the `sokoban` must find at least one empty square behind him in the direction `OppDir`. The `choose_move_destination` predicate returns an empty location behind the `sokoban`. This predicate is nondeterministic. Upon backtracking, this predicate returns another empty location behind the `sokoban`. After

```
import planner.

final({_SoLoc,BoxLocs}) =>
    foreach ((R,C) in BoxLocs)
       goal(R,C)
    end.

action({SoLoc,BoxLocs},NextS,Action,Cost) ?=>   % pull a box
    NextS = {NewSoLoc,NewBoxLocs},
    Action = $pull(BoxLoc,NewBoxLoc,OppDir), Cost=1,
    neib(SoLoc,BoxLoc,Dir),
    select(BoxLoc,BoxLocs,BoxLocs1),
    OppDir = opposite(Dir),
    neib(SoLoc,PrevLoc,OppDir),
    not member(PrevLoc,BoxLocs1),
    choose_move_destination(PrevLoc,BoxLocs1,OppDir,NewSoLoc),
    neib(NewSoLoc,NewBoxLoc,Dir),
    insert_ordered(BoxLocs1,NewBoxLoc) = NewBoxLocs.
action({SoLoc,BoxLocs},NextS,Action,Cost) =>    % walk
    NextS = {NewSoLoc,BoxLocs},
    Action = $walk(SoLoc,NewSoLoc,Dir), Cost=1,
    neib(SoLoc,NextLoc,Dir),
    not member(NextLoc,BoxLocs),
    choose_move_destination(NextLoc,BoxLocs,Dir,NewSoLoc).

opposite(up) = down.
opposite(down) = up.
opposite(left) = right.
opposite(right) = left.

table
neib((R,C),Next,Dir) =>
    Next = (R1,C1),
    Neibs = [(R-1,C,up),(R+1,C,down),(R,C-1,left),(R,C+1,right)],
    member((R1,C1,Dir),Neibs),
    size(NRows,NCols),
    R1 > 1, R1 < NRows,
    C1 > 1, C1 < NCols,
    not wall(R1,C1).

choose_move_destination(Loc,_BoxLocs,_Dir,Dest) ?=> Dest = Loc.
choose_move_destination(Loc,BoxLocs,Dir,Dest) =>
    neib(Loc,NextLoc,Dir),
    not member(NextLoc,BoxLocs),
    choose_move_destination(NextLoc,BoxLocs,Dir,Dest).
```

Fig. 5.6 A program for reversed Sokoban

pulling the box at BoxLoc to NewBoxLoc, the sokoban moves to NewSoLoc. The second rule of action/4 lets the sokoban walk to a new location.

For the reversed instance that is shown in Fig. 5.5, the program, using Picat version 1.3, finds an optimal plan of 88 steps in 0.2 s on an Intel-i5 PC. This plan can be converted to a plan of 86 steps for the original instance. The converted plan has two fewer steps, since the two walking moves are not necessary after the goal state has been achieved.

5.7 Bibliographical Note

PLANNER [25], which was designed as "a language for proving theorems and manipulating models in a robot", is perceived as the first logic programming language. Planning has been an important problem domain for Prolog [33, 62]. Planning is also an important application domain of answer set programming [9, 20, 35], which belongs to the planning-as-satisfiability approach [30]. The planner module of Picat is motivated by several successful applications of tabled logic programming for planning [5, 69, 71].

The Planning Domain Definition Language (PDDL) is the de facto standard planning language used by the planning community [39]. PDDL was inspired by STRIPS [19], and was made popular by the International Planning Competition. A detailed description of PDDL and its extensions, including the Hierarchical Task Network (HTN), can be found in [21].

As a modeling language for planning, Picat differs from PDDL and ASP in the following aspects: (1) Picat allows the use of structures to represent states; (2) Picat supports explicit commitment and nondeterministic actions, which enables users to have better control over action applications; (3) as presented in Chap. 6, Picat provides facilities for describing domain knowledge and heuristics for pruning search space. A PDDL encoding can be converted to Picat in a straightforward manner. Nevertheless, the resulting encoding can hardly be efficient. Picat's modeling and solving features need to be fully exploited in order to find efficient encodings [4, 70].

The idea to solve Sokoban by reversing the puzzle is discussed in [55].

5.8 Exercises

1. Write a program to solve the *water jugs problem*. There are an empty 3-gallon jug and an empty 5-gallon jug. The jugs can be filled from a fountain of water, water can be poured between the two jugs, and the jugs can be emptied. The problem is to fill one of the jugs with exactly 4 gallons of water.
2. Write a program to solve the *bridge-crossing problem*. Four people come to a river during the night. There is a narrow bridge, but it can only hold two people at a time. They have one torch and, because it's night, the torch has to be used when crossing the bridge. Person A can cross the bridge in one minute, person B in two minutes, person C in five minutes, and person D in eight minutes. When two people cross the bridge together, they must move at the slower person's pace.

Find an optimal plan that requires the shortest time for the four people to cross the bridge.

3. Write a program to solve the *wolf-chicken-feed problem*. A farmer must transport a wolf, a chicken, and a bag of chicken feed across a river. However, the boat can only hold two passengers, and the farmer is the only one who can row the boat across the river. (The bag of feed is also considered a passenger.) If the wolf is left alone with the chicken, then the wolf will eat the chicken. If the chicken is left alone with the bag of feed, then the chicken will eat the bag of feed. Find a plan to move them to the other side of the river.

4. Write a program to solve the *tile-swapping puzzle*.[1] The puzzle is a 3×3 board consisting of numbers from 1 to 9. Given an initial state, the objective of the puzzle is to swap the tiles until the goal state is reached. The following gives an initial state and the goal state:

```
7 3 2
4 1 5
6 8 9
```

```
1 2 3
4 5 6
7 8 9
```

Two adjacent tiles can be swapped if their sum is a prime number. Two tiles are considered adjacent if they have a common edge.

5. Write a program to solve the *moving coins problem* (see footnote 1). There is a line with 1000 cells numbered sequentially from 1 to 1000 from. N ($N < 1000$) coins are placed on the line. Coin i is placed at cell X_i, and no two coins are placed at the same cell. Bob would like to move the coins to the N leftmost cells of the line. To do this, he is allowed to take a coin from any cell T and move it to cell $T - j$, where j is an integer between 1 and K ($1 \leq K \leq 1000$), inclusive. This action is possible only if:

- cell $T - j$ actually exists and doesn't contain a coin;
- each of the cells $T - j + 1, \ldots, T - 1$ contains a coin.

Find a shortest plan for Bob to achieve his goal for two input integers N and K.

6. Write a program to solve the *white knight and black pawns problem*. There is a chessboard of size $N \times N$. A white knight and several black pawns are located on the board. The knight can move similarly to the normal knight in the game of chess. This means that the knight can either move (i) one horizontal space and two vertical spaces, or (ii) two horizontal spaces and one vertical space. The mission

---

[1]This problem is adapted from codechef.com.

Fig. 5.7 An unsolvable configuration in reverse Sokoban

of the knight is to capture as many black pawns as possible in a given number of moves.

7. Change the reverse Sokoban program shown in Fig. 5.6 as follows:

   (a) Modify the `main` predicate so that it accepts a description of a standard Sokoban instance, and then automatically converts the description to reverse Sokoban.

   (b) After a plan for reverse Sokoban is found, convert it to a plan for standard Sokoban, using pushes instead of pulls.

8. Although boxes in reverse Sokoban are less likely to get stuck than in standard Sokoban, the `sokoban` can get stuck. Figure 5.7 gives an unsolvable configuration in which the `sokoban` is stuck. Improve the program shown in Fig. 5.6 to avoid producing this kind of unsolvable configuration.

Chapter 6
Planning with Resource-Bounded Search

Abstract In addition to depth-unbounded search predicates, Picat's planner also provides *resource-bounded search* predicates. Resource-bounded search amounts to utilizing tabled states and their resource amounts to effectively decide when a state should be expanded and when a state should fail. Picat's planner provides *iterative-deepening* and *branch-and-bound search* predicates for finding optimal plans. Picat's planner also supports *heuristic search*. A state should fail if its heuristic estimate of the cost to reach a final state is greater than its resource amount. This chapter describes resource-bounded search predicates in the `planner` module, and demonstrates several application examples.

6.1 Introduction

For many planning problems, depth-unbounded depth-first search does not work if the search space is huge, with few dead-ends or few recurring states. Consider the *Deadfish problem*. Deadfish is a simple programming language that has a single accumulator to store data, and four instructions that modify the accumulator. The accumulator always starts with 0. The four instructions that modify the accumulator are: *i* (increment), *d* (decrement), *s* (square), and *o* (output). The Deadfish problem is to find the shortest sequence of instructions that produces a given non-negative number. Depth-unbounded search does not work for this problem since the search space is infinite, meaning that a path in the search space can go infinitely deep.

Picat's `planner` module provides predicates for performing *resource-bounded search*. In resource-bounded search, each state carries a resource amount. States and their resource amounts are tabled in such a way that a state is only expanded if: (i) the state is new and its resource amount is non-negative, or (ii) the state has previously failed, but the current occurrence has a higher resource amount than earlier occurrences. Resource-bounded search avoids unfruitful exploration of paths that are deemed to fail. *Depth-bounded search* is a special form of resource-bounded search in which plan lengths are limited.

In addition to the basic predicate that performs resource-bounded search, Picat's planner provides predicates that perform *iterative-deepening* and *branch-and-bound search* for finding optimal plans. Picat's planner also supports *heuristic search*. After a new state is generated, a decision can be made about whether to explore the state or fail the state, based on a heuristic value for the state and the current resource amount.

This chapter describes the resource-bounded search predicates in the `planner` module, together with their implementations and several application examples.

6.2 The `Planner` Module: Resource-Bounded Search

The `planner` module provides the following built-ins for resource-bounded and heuristic search:

- `plan` (*S*, *Limit*, *Plan*, *PlanCost*): This predicate searches for a plan by performing *resource-bounded search*. The predicate binds *Plan* to a plan that can transform state *S* to a final state that satisfies the condition given by `final/1`. *PlanCost* is the cost of *Plan*. *PlanCost* cannot exceed *Limit*, which is a given non-negative integer. The arguments *Limit* and *PlanCost* are optional.
- `best_plan` (*S*, *Limit*, *Plan*, *PlanCost*): This predicate finds an optimal plan by using *iterative-deepening*. The `best_plan` predicate calls the `plan/4` predicate to find a plan, using 0 as the initial cost limit and gradually relaxing the cost limit until a plan is found.
- `best_plan_bb` (*S*, *Limit*, *Plan*, *PlanCost*): This predicate finds an optimal plan by using *branch-and-bound*. First, the `best_plan_bb` predicate calls `plan/4` to find a plan. Then, it tries to find a better plan by imposing a stricter limit. This step is repeated until no better plan can be found. Then, this predicate returns the best plan that was found.
- `current_resource`() = *Limit*: This function returns the resource limit argument of the latest call to `plan/4`. In order to retrieve the *Limit* argument, the implementation has to traverse the call-stack until it reaches a call to `plan/4`. The `current_resource` function can be used to check against a heuristic value. If the heuristic estimate of the cost to travel from the current state to a final state is greater than the resource limit, then the current state should fail. If the estimated cost never exceeds the real cost, meaning that the heuristic function is *admissible*, then the optimality of solutions is guaranteed.

Solving tip: for a planning problem, the user has to decide whether to use depth-unbounded search or resource-bounded search, and if resource-bounded search is employed to find an optimal plan, then the user has to decide whether iterative-deepening or branch-and-bound should be used. In general, this decision requires experimentation with different search predicates.

6.3 The Implementation of Resource-Bounded Search

Figure 6.1 sketches Picat's implementation of the predicate `plan/4`. As in the implementation of `plan_unbounded`, the following array is passed from a state to the next state as an nt argument:

```
{Limit,IPlan,IPlanCost}
```

Unlike in the implementation of `plan_unbounded/4` that is shown in Fig. 5.1, where the nt argument is not tabled at all, Picat stores the `Limit` argument of each failed call to `plan_bounded_aux`,[1] and uses this information to decide whether the same state should fail when it recurs.

The first rule of `plan_bounded_aux` throws the inherited plan and its cost as an exception if `final(S)` succeeds. The exception will be caught by the `catch` call in the rule body of `plan/4`.

The second rule of `plan_bounded_aux` calls `action/4` to select an action, which produces a new state, NextS. Then, the rule computes the new resource limit, `Limit1`, by subtracting the cost of the selected action from `Limit`. If `Limit1 >= 0` succeeds, then the rule continues with the tabled search by recursively calling `plan_bounded_aux` on the new state NextS. Otherwise, if `Limit1 >= 0` fails, then Picat backtracks to select an alternative action.

The idea of resource-bounded search is to utilize tabled states and their resource limits to effectively decide when a state should be expanded and when a state should fail. Let S^R denote a state with an associated resource limit, R, as depicted in Fig. 6.2.

```
plan(S,Limit,Plan,PlanCost) =>
    IPlan = {Limit,[],0},
    catch(plan_bounded_aux(S,IPlan), (Plan,PlanCost), true).

table (+,nt)
plan_bounded_aux(S,{Limit,IPlan,IPlanCost}),
    final(S)
=>
    throw((IPlan.reverse(), IPlanCost)).
plan_bounded_aux(S,{Limit,IPlan,IPlanCost}) =>
    action(S,NextS,Action,ACost),
    Limit1 = Limit-ACost,
    Limit1 >= 0,
    Inherited1 = {Limit1,
                  [Action|IPlan],
                  IPlanCost+ACost},
    plan_bounded_aux(NextS,Inherited1).
```

Fig. 6.1 The implementation of `plan/4`

---

[1]This requires the implementation to treat the `Limit` argument of the predicate differently from other nt arguments.

Fig. 6.2 Resource-bounded search

If R is negative, then S^R immediately fails. If R is non-negative and S has never been encountered before, then S is expanded by using a selected action. Otherwise, if the same state S has failed before and R' was the resource limit when it failed, then S^R is only expanded if $R > R'$, i.e., if the current resource limit is larger than the resource limit was at the time of failure.

6.4 The Deadfish Problem

As described at the beginning of this chapter, the Deadfish problem is to find a shortest sequence of Deadfish instructions that produces a given non-negative number. Figure 6.3 gives a program for the problem.

```
import planner.

main =>
    print("Type a non-negative integer:"),
    N = read_int(),
    best_plan(((0,N),Plan),
    printf("%s\n", Plan ++ [o]),
    println(num_of_insts=len(Plan)+1).

final((N, N)) => true.

action((A,N),NextS,Action,Cost) =>
    NextS = (NextA,N), Cost = 1,
    instruction(A,N,NextA,Action).

instruction(A,N,NextA,Action), A < N ?=>
    Action = i, NextA = A+1.
instruction(A,N,NextA,Action), A !== 0 ?=>
    Action = d, NextA = A-1.
instruction(A,N,NextA,Action), A > 1, A < N =>
    Action = s, NextA = A*A.
```

Fig. 6.3 A program for the Deadfish problem

A state is represented as a pair, `(A,N)`, where `A` is the current value of the
accumulator, and `N` is the target number. The initial state is `(0,N)`, where `0` is the
initial value of the accumulator, and `N` is an input target value. The call `best_plan`
finds an optimal plan by using iterative deepening. Since the target number `N` can be
reached from 0 with at most `N` instructions, the following call, which uses branch-
and-bound, can be used instead:

```
best_plan((0,N),N,Plan).
```

The `final` predicate is defined as follows:

```
final((N, N)) => true.
```

A state is final if the accumulator has the same value as the target.

The `instruction` predicate defines the three instructions: `i` (increment), `d`
(decrement), and `s` (square). The `action` predicate calls `instruction` to apply
one of the instructions to the current value, `A`, to produce a new value, `NextA`

This program does not use any heuristic to estimate the number of instructions
that are needed to produce the target value from the current value. The `action`
predicate can be defined to incorporate a heuristic function as follows:

```
action((A,N),NextS,Action,Cost) =>
    NextS = (NextA,N),
    Cost = 1,
    instruction(A,NextA,Action),
    current_resource() > estimate_cost(NextA,N).
```

where `estimate_cost(NextA,N)` returns an estimate for the number of instructions that are needed to produce `N` from `NextA` See Exercise 1 for more details.

6.5 The 15-Puzzle

The *15-puzzle* consists of a 4×4 board and fifteen tiles numbered from 1 to 15. In
a configuration, each tile occupies a square on the board, and one square is empty.
The goal of the puzzle is to arrange the tiles from their initial configuration to a
goal configuration by making vertical or horizontal sliding moves that use the empty
square. Figure 6.4 shows an instance.

Figure 6.5 gives a program for the problem. A state is represented as a list of
sixteen locations, each of which takes the form (R_i, C_i), where R_i is a row number
and C_i is a column number. The first element in the list gives the position of the empty
square, and the remaining elements in the list give the positions of the numbered tiles
from 1 to 15.

The instance shown in Figure 6.4 can be solved by the following predicate:

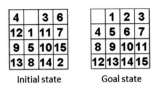

Initial state   Goal state

Fig. 6.4   15-Puzzle

```
import planner.

final(State) => State=[(1,1),(1,2),(1,3),(1,4),
                       (2,1),(2,2),(2,3),(2,4),
                       (3,1),(3,2),(3,3),(3,4),
                       (4,1),(4,2),(4,3),(4,4)].

action([P0@(R0,C0)|Tiles],NextS,Action,Cost) =>
    Cost = 1,
    (R1 = R0-1, R1 >= 1, C1 = C0, Action = up;
     R1 = R0+1, R1 =< 4, C1 = C0, Action = down;
     R1 = R0, C1 = C0-1, C1 >= 1, Action = left;
     R1 = R0, C1 = C0+1, C1 =< 4, Action = right),
    P1 = (R1,C1),
    slide(P0,P1,Tiles,NTiles),
    current_resource() > manhattan_dist(NTiles),
    NextS = [P1|NTiles].

% slide the tile at P1 to the empty square at P0
slide(P0,P1,[P1|Tiles],NTiles) =>
    NTiles = [P0|Tiles].
slide(P0,P1,[Tile|Tiles],NTiles) =>
    NTiles=[Tile|NTilesR],
    slide(P0,P1,Tiles,NTilesR).

manhattan_dist(Tiles) = Dist =>
    final([_|FTiles]),
    Dist = sum([abs(R-FR)+abs(C-FC) :
                {(R,C),(FR,FC)} in zip(Tiles,FTiles)]).
```

Fig. 6.5   A program for the 15-puzzle

```
main =>
    InitS = [(1,2),(2,2),(4,4),(1,3),
             (1,1),(3,2),(1,4),(2,4),
             (4,2),(3,1),(3,3),(2,3),
             (2,1),(4,1),(4,3),(3,4)],
    best_plan(InitS,Plan),
    foreach (Action in Plan)
       println(Action)
    end.
```

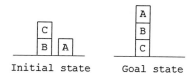

Fig. 6.6 Blocks World

In the initial state, the empty square is located at (1,2), tile #1 is located at (2,2), tile #2 is located at (4,4), and so on.

The final/1 predicate represents the configuration in which the empty square is located at (1,1), the upper-left corner of the board, and the numbered tiles are ordered from left to right and from top to bottom, as shown in Fig. 6.4.

The action/4 predicate is defined by one rule, which chooses one of four possible directions to move the empty square: up, down, left, and right. Let P0 be the current position of the empty square. After it is moved, its position is changed to P1. The slide predicate changes the current list of tile positions, Tiles, into NTiles by moving the tile that is currently at P1 to P0.

The test current_resource() > manhattan_dist(NTiles) ensures that the current resource limit is greater than the Manhattan distance from the tiles' current configuration to the tiles' final configuration. Note that the relation is >, and not ≥, because the current action costs 1, which has not yet been counted into the total cost.

Recall that the function zip(L_1,L_2) returns an association list of array pairs. For each tile, let (R,C) be the tile's current position, and let (FR,FC) be the tile's final position. The manhattan_dist(Tiles) function adds the distance abs(R-FR)+abs(C-FC) into the total cost.

The 15-puzzle program for the instance shown in Fig. 6.6, using Picat version 1.3, finds an optimal plan of 33 steps in 1.2 s on an Intel-i5 PC.

6.6 Blocks World

Blocks World is one of the most famous AI planning problems. As illustrated in Fig. 6.6, there are a given number of cubical blocks, a table that is large enough to hold all of the blocks, and a robot. Each block either sits on another block or sits on the table. A block is *clear* if there is no block sitting on it. The robot can pile a clear block onto another block or move it onto the table. Given an initial configuration and a goal configuration of blocks, the problem is to find a plan for the robot to transform the initial configuration to the goal configuration.

Figure 6.7 gives a program for the problem. A state is represented as a pair of configurations, (Piles,GPiles), where Piles represents the current configuration of the piles, and GPiles is the configuration of the piles in the goal state. A configuration of piles is represented as a sorted list of piles, where each pile is a list

```
import planner.

final((Piles,Piles)) => true.

% move a block from a pile to another pile
action((Piles,GPiles),NextS,Action,Cost),
     Cost = 1,
     NextS = (NPiles,GPiles),
     select(FromPile,Piles,Piles1),
     not stable(FromPile,GPiles),
     FromPile = [Block|FromPileR],
     select(ToPile,Piles1,PilesR),
     stable([Block|ToPile],GPiles)
=>
     NPiles1 = PilesR.insert_ordered([Block|ToPile]),
     if FromPileR == [] then
         NPiles = NPiles1
     else
         NPiles = NPiles1.insert_ordered(FromPileR)
     end,
     Action = $move(Block,ToPile).

% move a block from a pile to the table
action((Piles,GPiles),NextS,Action,Cost) =>
     Cost = 1,
     NextS = (NPiles,GPiles),
     select(FromPile,Piles,PilesR),
     not stable(FromPile,GPiles),
     FromPile = [Block|FromPileR],
     FromPileR !== [],
     NPiles1 = PilesR.insert_ordered([Block]),
     NPiles = NPiles1.insert_ordered(FromPileR),
     Action = $move(Block,table).

stable(Pile,GPiles),
     member(GPile,GPiles),
     append(_,Pile,GPile)
=>
     true.
```

Fig. 6.7 A program for Blocks World

of blocks in which the first element represents the block that is on the top of the pile.
The goal configuration is carried as part of each state in order to facilitate testing
against the final state and the utilization of domain knowledge.

The instance in Fig. 6.6 can be solved by the following predicate:

```
main =>
     Init = [[a],[c,b]],
     Goal = [[a,b,c]],
```

```
best_plan_upward((Init,Goal),Plan),
foreach (Step in Plan)
   println(Step)
end.
```

The initial configuration consists of two piles, [a] and [c,b]. The goal configuration consists of a single pile, [a,b,c]. Note that, in both the initial configuration and the final configuration, the list of piles must be sorted. For example, if the initial state were given as Init = [[c,b],[a]], then the program would not work. The program uses the branch-and-bound search predicate best_plan_upward, which works better than iterative-deepening for this problem. Since both action rules are deterministic, the program only returns one plan, which is guaranteed to be optimal. For this reason, best_plan_upward can be replaced by plan.

The final/1 predicate is defined as follows:

```
final((Piles,Piles)) => true.
```

A state is final if the current configuration is the same as the goal configuration.

The action/4 predicate is defined by two rules. The first rule moves a clear block from a pile to another pile. The call select(FromPile, Piles, Piles1) selects an origin pile, FromPile, from Piles, returning the remaining list of piles as Piles1. The call select(ToPile, Piles1, PilesR) selects a destination pile, ToPile, from Piles1, returning the remaining list of piles as PilesR. This rule moves the Block from the top of FromPile to the top of ToPile. This rule not only represents the pre-conditions and effects of the move action, but also incorporates the following domain knowledge: a pile cannot be destroyed if it is already *stable* in the goal configuration, and a block cannot be stacked onto a pile unless the resulting pile is stable in the goal configuration. The second action rule moves a clear block from a pile onto the table. The stable predicate tests if a pile is stable. A pile is stable if it is the suffix of a pile in the goal configuration.

The Blocks World program is very efficient, since both action rules are deterministic. This program becomes more difficult if the number of piles that can be held by the table is limited. For more details, see Exercise 4.

6.7 Logistics Planning

Many military and business operations require *logistics planning*, which includes the ability to efficiently transport people and supplies. This section considers the *Transport domain* taken from the International Planning Competition 2014 (IPC'14). There is a weighted directed graph, a set of trucks, each of which has a capacity that indicates the maximum number of packages that the truck can carry, and a set of packages, each of which has an initial location and a destination. There are three types of actions: *load* a package onto a truck, *unload* a package from a truck, and *move* a truck from one location to a different location. Each action has an associated cost. The objective of the problem is to find an optimal, minimum-cost plan to transport the packages from their initial locations to their destinations. The weighted directed graph is given by the predicate road(From,To,Cost), which succeeds if there is an edge from node From to node To that has a cost of Cost. An optimal plan for this problem normally requires trucks to cooperate. This problem degenerates into the *shortest path problem* if there is only one truck and only one package.

Figure 6.8 gives a program for the problem. A state is represented as an array of the form {Trucks,Packages}, where Trucks is an ordered list of trucks, and Packages is an ordered list of waiting packages. A package in Packages is a pair of the form (Loc,Dest), where Loc is the source location of the package, and Dest is the destination of the package. A truck in Trucks is a list of the form [Loc,Dests,Cap], where Loc is the current location of the truck, Dests is an ordered list of destinations of the packages that are loaded onto the truck, and Cap is the capacity of the truck. Only the destination of a loaded package is included in the package's representation; once a package is loaded, its origin is irrelevant for planning, and its location is the same as the truck's location. At all times, the number of loaded packages must not exceed the truck's capacity.

Note that keeping the capacity as the last element of the list facilitates sharing, since the suffix [Cap], which is common to all the trucks that have the same capacity, is only tabled once. Also note that the names of the trucks and the names of the packages are not included in the representation. The representation breaks symmetries as follows: If two packages have the same destination, then they are indistinguishable if: (i) they are both in the waiting list, and both have the same source, or (ii) they are both on the same truck. Furthermore, two configurations are treated as the same state if they only differ by a truck's name or a package's name.

The following predicate solves an instance:

```
main =>
    Facts =
        $[road(c3,c1,40),road(c1,c3,40),road(c3,c2,18),
          road(c2,c3,18),road(c4,c1,36),road(c1,c4,36),
          road(c4,c3,37),road(c3,c4,37),road(c5,c2,24),
          road(c2,c5,24),road(c5,c3,26),road(c3,c5,26)],
    cl_facts(Facts,[$road(+,-,-)]),
    Trucks = [[c1,[],2],[c2,[],3]],
    Packages = [(c1,c2),(c1,c2),(c2,c5),(c3,c1)],
    best_plan({Trucks,Packages},Plan),
    foreach (Action in Plan)
        println(Action)
    end.
```

```
import planner.

final({Trucks,[]}) =>
    foreach([_Loc,Dests|_] in Trucks)
        Dests == []
    end.

action({Trucks,Packages},NextState,Action,ActionCost) ?=>
    Action = $unload(Loc),
    ActionCost = 1,
    select([Loc,Dests,Cap],Trucks,TrucksR),
    select(Dest,Dests,DestsR),
    NewTrucks = insert_ordered(TrucksR,[Loc,DestsR,Cap]),
    (Loc == Dest ->
        NewPackages = Packages;
        NewPackages = insert_ordered(Packages,(Loc,Dest))),
    NextState = {NewTrucks,NewPackages}.
action({Trucks,Packages},NextState,Action,ActionCost) ?=>
    Action = $load(Loc),
    ActionCost = 1,
    select([Loc,Dests,Cap],Trucks,TrucksR),
    length(Dests) < Cap,
    select((Loc,Dest),Packages,PackagesR),
    NewDests = insert_ordered(Dests,Dest),
    NewTrucks = insert_ordered(TrucksR,[Loc,NewDests,Cap]),
    NextState = {NewTrucks,PackagesR}.
action({Trucks,Packages},NextState,Action,ActionCost) =>
    Action = $move(Loc,NextLoc),
    select([Loc|Tail],Trucks,TrucksR),
    road(Loc,NextLoc,ActionCost),
    NewTrucks = insert_ordered(TrucksR,[NextLoc|Tail]),
    NextState = {NewTrucks,Packages},
    estimate_cost(NextState) =< current_resource()-ActionCost.

table
estimate_cost({Trucks,Packages}) = Cost =>
    LoadedPackages = [(Loc,Dest) :
                        [Loc,Dests,_] in Trucks, Dest in Dests],
    NumLoadedPackages = length(LoadedPackages),
    TruckLocs = [Loc : [Loc|_] in Trucks],
    travel_cost(TruckLocs,LoadedPackages,Packages,0,TCost),
    Cost = TCost+NumLoadedPackages+length(Packages)*2.

% the maximum of the minimum costs
travel_cost(_Trucks,[],[],Cost0,Cost) => Cost=Cost0.
travel_cost(Trucks,[(PLoc,PDest)|Packages],Packages2,Cost0,Cost) =>
    Cost1 = min([D1+D2 : TLoc in Trucks,
                        shortest_dist(TLoc,PLoc,D1),
                        shortest_dist(PLoc,PDest,D2)]),
    travel_cost(Trucks,Packages,Packages2,max(Cost0,Cost1),Cost).
travel_cost(Trucks,[],Packages2,Cost0,Cost) =>
    travel_cost(Trucks,Packages2,[],Cost0,Cost).

table (+,+,min)
shortest_dist(X,X,Dist) => Dist=0.
shortest_dist(X,Y,Dist) =>
    road(X,Z,DistXZ),
    shortest_dist(Z,Y,DistZY),
    Dist = DistXZ+DistZY.
```

Fig. 6.8 A program for the Transport domain

Note that both `Trucks` and `Packages` must be sorted.

The `final/1` predicate is defined as follows:

```
final({Trucks,[]}) =>
    foreach([_Loc,Dests|_] in Trucks)
        Dests == []
    end.
```

A state is final if all of the trucks are empty, and if there are no waiting packages.

The `action/4` predicate is defined by three rules, each defining an action type. For the *load* action, the rule nondeterministically selects a truck that still has room for another package, and nondeterministically selects a package that has the same location as the truck. After loading the package onto the truck, the rule inserts the package's destination into the list of the truck's loaded packages, if the truck's location is not the same as the package's destination. Note that the rule is nondeterministic. Even if a truck passes by a location that has a waiting package, the truck may not take the package. If this rule is made deterministic, then the optimality of plans is no longer guaranteed, unless there is only one truck, whose capacity is infinite.

The program can be improved by incorporating some domain control knowledge. The following rule can be inserted into the beginning of `action/4` to reduce non-determinism:

```
action({Trucks,Packages},NextS,Action,ACost),
    select([Loc,Dests,Cap],Trucks,TrucksR),
    select(Loc,Dests,DestsR)
=>
    Action = $unload(Loc),
    ACost = 1,
    NewTrucks = insert_ordered(TrucksR,[Loc,DestsR,Cap]),
    NextS = {NewTrucks,Packages}.
```

This rule deterministically unloads a package if the package's destination is the same as the truck's location.

After each new state is generated, the following condition is checked to ensure that the current path is viable:

```
    current_resource() - ACost >=  estimate_cost(NextS).
```

Let $P_1, \ldots, P_n$ be the remaining packages, and let C_i be the minimum cost of moving package P_i to its destination. The moving cost of a state can safely be estimated as $\max(\{C_1, \ldots, C_n\})$. The estimated total cost is the estimated moving cost plus the loading and unloading costs of all of the remaining packages. This heuristic is a lower bound of the real cost, and is therefore admissible. The cost of a minimum Steiner tree that connects the locations of the packages to their destinations is a more accurate estimate of the moving cost. For more details, see Exercise 5.

6.8 Bibliographical Note

The iterative-deepening A* algorithm, IDA*, was first proposed by Richard E. Korf in [32]. The IDA* algorithm and many other algorithms that are used by planning systems can be found in [34, 46]. The resource-bounded tabled search was detailed in [70]. Unlike IDA*, which starts a new search round from scratch, the `best_plan` search predicate reuses the states that were tabled in the previous rounds.

In the past, a lot of work was done on the use of domain knowledge in planning [2, 21, 23, 31], but recently this part of modeling has been put aside, because of the advancement of domain-independent PDDL planners. It has been shown that with structural state representations that facilitate term sharing and symmetry breaking, and with domain-dependent control knowledge and heuristics, Picat models can be significantly faster than PDDL models that are solved by the best domain-independent planners [4, 70].

6.9 Exercises

1. The Deadfish program shown in Fig. 6.3 does not use any heuristic. Let N be the target value, and let A be the accumulator's value. A heuristic function can compute the number of increment or decrement instructions that are needed to produce N or $\sqrt{N}$ from A if $\sqrt{N} < A < N$. Write such a heuristic function, and use it to make the Deadfish program faster.
2. The Sokoban problem from Sect. 5.6 was solved by using depth-unbounded search. This problem can also be solved by using the depth-bounded search predicate `best_plan.` Analyze the reason why depth-unbounded search is faster than depth-bounded search for the Sokoban problem. What heuristics can be utilized to make depth-bounded search more effective than depth-unbounded search for Sokoban?
3. The Manhattan-distance heuristic used in the 15-puzzle program in Sect. 6.5 computes the shortest distance of each tile from its current position to its final position and sums all the distances. This heuristic function is admissible but not accurate. Improve the 15-puzzle program by using a more effective heuristic function.
4. The program for Blocks World that is shown in Fig. 6.7 assumes that the table is big enough to hold any number of piles. Assume that the table is only big enough to hold three piles. Then, a block may have to be moved onto a pile even if the resulting pile is not stable. Rewrite the program to meet this new constraint.
5. The heuristic function that is used in the Transport program in Fig. 6.8 computes the maximum of the shortest distances of the packages from their current positions to their destinations. A more effective heuristic function can use the cost of a minimum Steiner tree as an estimate. Write a heuristic function to compute the cost of such a minimum Steiner tree, and investigate if it can improve the speed of the program.

6. Write a program to solve the *Tower of Hanoi problem*.[2] There are three pegs, and N disks of differing sizes that are stacked in increasing size order on the leftmost peg, with the smallest disk on top. N is input by the user. The goal is to find the shortest-length plan to stack the N disks on the rightmost peg, using the following rules:

   a. Only one disk can be moved at a time.
   b. Only the top disk on any peg can be moved.
   c. Larger disks cannot be stacked above smaller disks.

7. Modify the Tower of Hanoi program from Exercise 6 so that it can accept an arbitrary number of pegs, K, as input, and generate the shortest-length plan to stack the N disks on the rightmost peg.

---

[2]For a dynamic programming model of the problem, see Sect. 4.8.

Chapter 7
Encodings for the Traveling Salesman Problem

Abstract So far, this book has presented several techniques for modeling constraint satisfaction, dynamic programming, and planning problems. This chapter presents encodings for the Traveling Salesman Problem, comparing models for several solvers, including CP, SAT, MIP, and tabled planning.

7.1 Introduction

The *Traveling Salesman Problem* (TSP) is one of the most well-known combinatorial problems. Given a set of cities, the objective is to find a tour of all of the cities such that the total traveling cost is minimized. This problem has many applications, such as creating optimal computer boards, logistics, and telecommunications.

TSP is a graph problem. For a given graph, which can be directed or undirected, the goal of TSP is to find a *Hamiltonian cycle* that connects all of the vertices of the graph such that the total travel cost is minimized. This chapter presents encodings for the Traveling Salesman Problem for several solvers, including CP, SAT, MIP, and tabled planning.

7.2 An Encoding for CP

A well-known representation of a graph in CP is to use a domain variable for each vertex in the graph, where the domain is a set of neighboring vertices. For example, the graph shown in Fig. 7.1 can be represented by using six domain variables: V_1 for vertex 1, V_2 for vertex 2, and so on. Vertex 1 is directly connected to vertices 2, 3, and 4, so the domain of V_1 is [2,3,4]. Domain variables are not sufficient for representing weighted graphs; they need to be supplemented with a data structure for representing weights, such as a *cost matrix* in which the entry at (i, j) stores the weight of the edge between V_i and V_j.

© The Author(s) 2015
N.-F. Zhou et al., *Constraint Solving and Planning with Picat*,
SpringerBriefs in Intelligent Systems, DOI 10.1007/978-3-319-25883-6_7

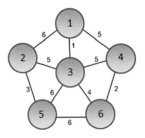

Fig. 7.1 A sample graph

The following shows a CP encoding for TSP:

```
import cp.

main =>
    M = {{0,6,1,5,0,0},              % cost matrix
         {6,0,5,0,3,0},
         {1,5,0,5,6,4},
         {5,0,5,0,0,2},
         {0,3,6,0,0,6},
         {0,0,4,2,6,0}},
    tsp(M).

tsp(M) =>
    N = length(M),
    NextArr = new_array(N),          % visit NextArr[I] after I
    NextArr :: 1..N,
    CostArr = new_array(N),
    circuit(NextArr),
    foreach (I in 1..N)
       CostArr[I] #> 0,
       element(NextArr[I],M[I],CostArr[I])
    end,
    TotalCost #= sum(CostArr),
    solve($[min(TotalCost), report(println(cost=TotalCost))],
          NextArr),
    foreach (I in 1..N)
       printf("%w -> %w\n",I,NextArr[I])
    end.
```

The tsp/1 predicate takes the cost matrix of a graph and searches for an optimal route over the graph. The function length(M) returns the size of M, which is the same as the number of vertices in the graph. The NextArr array serves as a holder for a route to be found. For each vertex `I`, the successor vertex of `I` in the route is `NextArr[I]`. The call `NextArr :: 1..N` narrows the domain of each of the variables in `NextArr` to the *range* `1..N`. After this call, the variables in `NextArr` become *domain* variables, which can only take values from the domain.

The call `CostArr = new_array(N)` binds `CostArr` to another N-element array. This array represents the costs of the edges in the route. This call is followed by a call to the global constraint `circuit`,[1] which is made precisely for the purpose of connecting each vertex with the next vertex to visit, with the additional constraint that all vertices must be visited exactly once. The `circuit` constraint enforces that it is a *tour*, meaning that after the last vertex is visited, the first vertex must be visited.

For each vertex `I`, let `J` be the successor of `I` in the route, i.e.,

```
J = NextArr[I].
```

Then `CostArr[I]` is the same as `M[I,J]`, the cost of the edge `I→J`. Since array indices must be integers in Picat, it is impossible to write the constraint as `CostArr[I] = M[I,J]`, because `J` is a variable. For this reason, the `element`[2] constraint is used to force `CostArr[I]` to be equal to the travel cost of the edge `I→J`.

The call `solve(...,NextArr)` starts the search by the CP solver. The first argument specifies a list of options to be passed to the solver. The dollar sign `$` informs the compiler that the elements in the list are terms, not function calls. The option `min(TotalCost)` specifies an objective value to be minimized. The option `report` prints out the current total cost each time a better route is found while searching for an optimal answer. Note that, since the variables in `CostArr` are functionally dependent on the variables in `NextArr`, it is unnecessary to pass `CostArr` to `solve`.

The call `solve` uses the default strategy for labeling variables with values, i.e., choosing variables from left to right, and selecting values from the minimum to the maximum. For the objective value `min(TotalCost)`, the CP solver employs branch-and-bound to find a solution that has the minimal value for `TotalCost`. The solver first searches for an assignment for `NextArr`, which gives a value C for `TotalCost`. Then, the solver restarts the search from scratch, imposing the constraint `TotalCost #< C`. These two steps are repeated until no solution can be found. At the end, the solver returns the last solution that was found.

A Better Labeling Strategy

For TSP, the default value-selecting strategy that selects values from the minimum to the maximum is not meaningful, since the vertices' numbers do not reflect the graph structure. One value-selecting strategy that is often used in greedy algorithms for TSP is to visit the neighbors by starting from the nearest and ending with the farthest. This strategy has the promise to start with a short tour that leads to more effective pruning

---

[1] http://sofdem.github.io/gccat/gccat/Ccircuit.html.

[2] http://sofdem.github.io/gccat/gccat/Celement.html.

during branch-and-bound search. The following shows an improved program that orders neighbors by distance, and selects the nearest neighbor first:

```
tsp(M) =>
    N = length(M),
    NextArr = new_array(N),      % visit NextArr[I] after I
    NextArr :: 1..N,
    CostArr = new_array(N),
    circuit(NextArr),
    foreach (I in 1..N)
       CostArr[I] #> 0,
       element(NextArr[I],M[I],CostArr[I])
    end,
    TotalCost #= sum(CostArr),
    foreach (I in 1..N)
       Pairs = [(M[I,J],J) : J in 1..N, M[I,J] > 0].sort(),
       Neibs = [J : (_,J) in Pairs],
       NextArr[I].put(domain,Neibs)
    end,
    solve($[min(TotalCost), label(mylabel)], NextArr),
    foreach (I in 1..N)
       printf("%w -> %w\n",I,NextArr[I])
    end.

mylabel(V),var(V) =>
    Domain = V.get(domain),
    member(V,Domain).
mylabel(_V) => true.
```

Before solve, a foreach is inserted to order the neighbors of each vertex. For each vertex I, the loop computes a sorted list of neighbors Neibs, and attaches Neibs to variable NextArr[I] as an attribute with the name domain.[3] The list comprehension [(M[I,J],J) : J in 1..N, M[I,J] > 0] returns a list of pairs of the form (M[I,J],J), where J is a neighbor of I, and M[I,J] is the cost that is associated with the edge I→J. The list comprehension [J : (_,J) in Pairs] extracts the neighbors from Pairs. The call NextArr[I].put(domain,Neibs) attaches Neibs as an attribute with the name domain to variable NextArr[I].

The option label(mylabel)[4] informs the CP solver that once a variable V is selected, the user-defined call mylabel(V) is used to label V. If V is a variable, then V.get(domain) retrieves the attribute value of domain attached to V, and

---

[3] A variable is called an *attributed variable* if one or more attributes are attached to it. A domain variable is a special kind of an attributed variable.

[4] The option label is supported in Picat version 1.4 and later.

the call member(V, Domain) nondeterministically assigns a value of Domain to V. Since the list Domain is sorted by distance, the neighboring vertices in the list are visited from the nearest to the farthest during search.

7.3 An Encoding for SAT

Since Picat provides a common interface to CP, SAT, and MIP solvers,[5] the above encodings can be made to work with SAT by changing import cp to import sat. Nevertheless, a model that works best for CP is not necessarily the best model for SAT.

In general, 0/1 integer programming models work better for the SAT solver than the CP solver. In the above CP models, a variable is used for each vertex, and the value for the variable indicates the next vertex to be visited. This model requires N variables. A 0/1 integer programming model uses an N×N matrix of Boolean variables: the entry at [I, J] is 1 iff vertex J is visited immediately after vertex I. The following Picat program encodes this model:

```
import sat.

tsp(M)  =>
    N = length(M),
    NextArr = new_array(N,N),
    NextArr :: 0..1,
    foreach (I in 1..N, J in 1..N)
       if M[I,J] == 0 then NextArr[I,J] = 0 end
    end,
    % each vertex is followed by exactly one vertex in a tour
    foreach (I in 1..N)
       sum([NextArr[I,J] : J in 1..N]) #= 1
    end,
    % each vertex is preceded by exactly one vertex in a tour
    foreach (J in 1..N)
       sum([NextArr[I,J] : I in 1..N]) #= 1
    end,
    % no sub-tours
    Order = new_array(N),
    Order :: 1..N,
    Order[1] = 1,       % visit vertex 1 first
    foreach (I in 1..N)
       NextArr[I,1] #=> Order[I] #= N,
       foreach (J in 2..N)
```

---

[5] In Picat version 1.3, the mip module only supports linear constraints.

```
              NextArr[I,J] #=> Order[J] #= Order[I]+1
          end
      end,
      CostArr = new_array(N),
      foreach (I in 1..N)
          CostArr[I] :: min([M[I,J] : J in 1..N, M[I,J] !== 0]) ..
                        max([M[I,J] : J in 1..N, M[I,J] !== 0]),
          foreach (J in 1..N)
              NextArr[I,J] #=> CostArr[I] #= M[I,J]
          end
      end,
      TotalCost #= sum([CostArr[I] : I in 1..N]),
      solve($[min(TotalCost),report(println(cost=TotalCost))],
          NextArr),
      foreach (I in 1..N, J in 1..N)
          if NextArr[I,J]==1 then printf("%w -> %w\n",I,J) end
      end.
```

NextArr is an N×N two-dimensional Boolean array. For two vertices I and J, a tour includes the edge I→J iff NextArr[I,J] is 1. The first foreach loop ensures that the edge I→J is not in a tour if vertex J is not directly reachable from vertex I or if J == I. The second foreach loop ensures that each vertex is followed by exactly one vertex in a tour. The third foreach loop ensures that each vertex is preceded by exactly one vertex in a tour.

In order to prevent sub-tours in solutions, this model assigns a unique ordering number from 1..N to each vertex. For each vertex I, OrderArr[I] indicates the ordering number assigned to vertex I. Without loss of generality, 1 is assigned to vertex 1, meaning that vertex 1 is visited first. For each vertex I, if vertex 1 is visited after vertex I, then OrderArr[I] equals N. Otherwise, for each J in 2..N, if NextArr[I,J] == 1, then OrderArr[J] #= OrderArr[I]+1.

As in the CP encoding, each vertex is assigned a cost. For each vertex I, let J be the successor of I in the route, i.e., NextArr[I,J] is 1. Then CostArr[I] is the same as M[I,J], the cost of the edge I→J.

When the sat module is used, constraints are compiled into a logic formula in the conjunctive normal form (CNF) for the underlying SAT solver. Picat employs the so called *log-encoding* for compiling domain variables and constraints. For a non-Boolean domain variable, $\lceil log_2(n) \rceil$ Boolean variables are used, where n is the maximum absolute value of the domain. If the domain contains both negative and positive values, then another Boolean variable is used to encode the sign. Each combination of values of these Boolean variables represents a valuation for the domain variable. If there are holes in the domain, then disequality (#!=) constraints are generated in order to disallow assignments of those hole values to the variable. Compared with other encodings such as sparse encoding and order encoding, log-encoding is especially suited to arithmetic constraints, because of the compactness of the generated code.

7.4 An Encoding for MIP

The above models contain non-linear constraints. In order to make these models work with the MIP solver, the compiler must linearize the non-linear constraints before sending them to the underlying MIP solver. The following Picat program encodes a linear integer programming model for the MIP solver:

```
import mip.

tsp(M) =>
    N = length(M),
    NextArr = new_array(N,N),
    NextArr :: 0..1,
    foreach (I in 1..N, J in 1..N)
       if M[I,J] == 0 then NextArr[I,J] = 0 end
    end,
    % each vertex is followed by exactly one vertex in a tour
    foreach (I in 1..N)
       sum([NextArr[I,J] : J in 1..N]) #= 1
    end,
    % each vertex is preceded by exactly one vertex in a tour
    foreach (J in 1..N)
       sum([NextArr[I,J] : I in 1..N]) #= 1
    end,
    % no sub-tours
    OrderArr = new_array(N),
    OrderArr :: 1..N,
    OrderArr[1] = 1,      % visit vertex 1 first
    % linear ordering constraints
    foreach (I in 2..N, J in 2..N, I !== J)
       OrderArr[I] - OrderArr[J] + NextArr[I,J]*N #=< N-1
    end,
    TotalCost #= sum([NextArr[I,J]*M[I,J] :
                      I in 1..N, J in 1..N, I !== J]),
    solve($[min(TotalCost)], NextArr),
    foreach (I in 1..N, J in 1..N)
       if NextArr[I,J]==1 then printf("%w -> %w\n",I,J) end
    end,
    println(cost=TotalCost).
```

This model is the same as the one for SAT, except that N^2 linear constraints are used to constrain the ordering variables in OrderArr. For each pair of vertices [I,J], where I !== J, the constraint

```
OrderArr[I] - OrderArr[J] + NextArr[I,J]*N #=< N-1
```

holds if NextArr[I,J] = 0. If NextArr[I,J] = 1, the constraint still holds
if the ordering number that is assigned to vertex J is greater than that assigned to
vertex I. It has been proven that these linear constraints guarantee the soundness and
completeness of the model: every solution that satisfies the constraints is guaranteed
not to contain any sub-tours; for every feasible tour, there is an assignment for the
ordering variables that satisfies the constraints.

7.5 An Encoding for the Tabled Planner

This section formulates TSP as a planning problem for Picat's planner. For a planning
problem, users only need to define the predicates final/1 and action/4, and
call one of the search predicates in the module on an initial state in order to find a
plan or an optimal plan.

For TSP, a state can be represented by a tuple of the form:

```
{CurrV,Unvisited}
```

where CurrV is the vertex that is being visited, and Unvisited is a sorted list of
vertices that have not yet been visited. The initial state is {1,2..N}, where 1 is the
current vertex that is being visited, and 2..N is a list of vertices to be visited.

The following encodes a planner model for TSP:

```
import planner.

tsp(M) =>
    N = length(M),
    Rel = [],
    foreach (I in 1..N)
        Neibs = [(J,M[I,J]) : J in 1..N, M[I,J] > 0].sort(2),
        Rel := [$neibs(I,Neibs)|Rel]
    end,
    cl_facts(Rel,[$neibs(+,-)]),
    best_plan_bb({1,2..N},Plan,Cost),
    println(Plan),
    println(plan_cost = Cost).

final({1,[]}) => true.

action({V,[]}, NState, Action, ActionCost) =>
    NState = {1,[]},
```

```
    Action = 1,                      % go back to vertex 1
    neibs(V,Neibs),
    member((1,ActionCost),Neibs).
action({V,Unvisited}, NState, Action, ActionCost) =>
    NState = {NextV,NUnvisited},
    Action = NextV,                  % visit NextV
    neibs(V,Neibs),
    member((NextV,ActionCost),Neibs),
    once select(NextV,Unvisited,NUnvisited).
```

The predicate `tsp(M)` first converts the given cost matrix into an adjacency relation. In the beginning, the relation `Rel` is empty. For each vertex `I`, the assignment `Rel := [$neibs(I,Neibs)|Rel]` adds the fact `neibs(I,Neibs)` into the relation, where `Neibs` is a sorted list of neighbor-distance pairs. Note that the list `Neibs` is sorted by the key index 2, i.e., the distance from vertex `I` to the neighbor. After the `foreach` loop, `cl_facts(Rel,[$neibs(+,-)])` compiles and loads the relation into the code area, assuming that the relation is indexed on the first argument.

The call `best_plan_bb` starts branch-and-bound search for a plan, starting with the initial state `{1, 2..N}`. In addition to branch-and-bound search, Picat also provides predicates for performing depth-unbounded search and iterative-deepening search.

The predicate `final/1` is defined by one rule. A state is final if it is `{1, []}`, which means that the vertex that is being visited is 1, and the list of vertices remaining to be visited is empty.

The predicate `action/4` is defined by two rules. The first rule states that if there are no remaining vertices to be visited, then the salesman must return to the starting vertex, i.e., vertex 1. The second rule handles the case in which the list of unvisited vertices is not empty. It selects a neighboring vertex, `NextV`, of `CurrV`, and removes `NextV` from `Unvisited`, resulting in a new list of unvisited vertices, `UnvisitedR`.

7.6 Experimental Results

Table 7.1 gives the costs of the solutions that are obtained by the four encodings for a set of randomly-generated TSP instances. For each instance name n_m_k, n is the number of vertices in the graph, m is the number of edges, and k is an instance number. Branch-and-bound is used in all of the solvers for finding optimal solutions. For MIP, a user-defined branch-and-bound is used so that the solver reports whenever a better solution is found. The solutions were obtained on a PC notebook with 2.4 GHz Intel i5 and 8 GB RAM, and the time limit was set to 10 min. None of the solvers succeeded in proving the optimality of any of the solutions.

Table 7.1 A comparison of TSP encodings on the costs of solutions

Benchmark	CP	SAT	MIP	Planner
70_300_0	380	443	765	n/a
70_300_11	471	509	708	n/a
70_300_12	403	433	645	n/a
70_300_14	407	499	n/a	n/a
70_300_3	482	518	739	n/a
70_300_4	487	532	n/a	n/a
70_300_5	424	503	672	417
70_300_7	465	501	658	n/a
70_300_8	453	497	738	n/a
70_300_9	420	472	619	n/a
80_340_0	507	572	n/a	n/a
80_340_10	535	521	847	n/a
80_340_11	541	656	n/a	n/a
80_340_13	476	530	n/a	n/a
80_340_15	516	516	n/a	n/a
80_340_16	580	714	n/a	n/a
80_340_17	502	607	n/a	n/a
80_340_18	469	576	n/a	n/a
80_340_4	539	575	n/a	n/a

CP is a clear winner. It found the best-quality solutions for all of the instances, except for instance 70_300_5. For CP, the labeling strategy that steers search toward low-cost solutions is effective.

SAT found at least one solution for each of the instances. The qualities of the solutions found by SAT are not as good as those found by CP. TSP is an arithmetic-intensive benchmark, and SAT is known to be weak at handling arithmetic constraints. The log-encoding of domain variables and constraints, which is adopted by the Picat SAT compiler, is less susceptible to space explosion than other encodings. Nevertheless, once constraints are compiled into CNF, the SAT solver is totally blind to the objective function, and cannot steer search toward low-cost solutions.

MIP failed to find any solutions for several of the instances, and for the instances for which it did find solutions, the qualities are not comparable with those found by CP or SAT. GLPK, the MIP solver that is linked to Picat, is not considered to be one of the fastest MIP solvers. The results could have been completely different had a cutting-edge MIP solver been used.

The planner encoding performed the worst; it failed to find solutions for any of of the instances, except for instance 70_300_5. Interestingly, the solution that was found by the planner for instance 70_300_5 turned out to be the best among the solutions that were found by any of the solvers. In the planner encoding, a state stores the current vertex that is being visited, and also a set of unvisited vertices. In the worst case, all subsets of the vertices need to be explored. Despite the potential memory blowup, all of the runs were terminated by out-of-time, and not by out-of-memory.

7.7 Bibliographical Note

TSP is an intensely investigated problem in computer science. Successful algorithms can find optimal solutions for graphs with over 10,000 vertices. For example, see [1] for a survey of methods for solving TSP. It is also one of the most well-studied problems in mathematical programming. For example, see [13] for integer programming models and techniques that are tailored to TSP. The `circuit` constraint [59] is motivated by graph problems that require finding Hamiltonian cycles, including TSP.

References

1. David L. Applegate, Robert E. Bixby, Vašek Chvátal, and William J. Cook. *The Traveling Salesman Problem: A Computational Study*. Princeton University Press, 2006.
2. Fahiem Bacchus and Froduald Kabanza. Using temporal logics to express search control knowledge for planning. *Artificial Intelligence*, 116(1–2):123–191, January 2000.
3. Roman Barták. Practical constraints: A tutorial on modelling with constraints. In *Proceedings of the 5th Workshop on Constraint Programming in Decision and Control*, 2003.
4. Roman Barták, Agostino Dovier, and Neng-Fa Zhou. On modeling planning problems in tabled logic programming. In *Proceedings of the 17th International Symposium on Principles and Practice of Declarative Programming*, pages 31–42, 2015.
5. Roman Bartak and Neng-Fa Zhou. Using tabled logic programming to solve the Petrobras planning problem. *Theory and Practice of Logic Programming, Special Issue on ICLP 2014*, 14(4–5):697–710, 2014.
6. Nicolas Beldiceanu, Mats Carlsson, and Jean-Xavier Rampon. The global constraint catalog. http://sofdem.github.io/gccat/.
7. Sally C. Brailsford, Chris N. Potts, and Barbara M. Smith. Constraint satisfaction problems: Algorithms and applications. *European Journal of Operational Research*, 119(3):557–581, December 1999.
8. Ivan Bratko. *Prolog for Artificial Intelligence (4th Edition)*. Addison-Wesley, 2011.
9. Gerhard Brewka, Thomas Eiter, and Miroslaw Truszczyński. Answer set programming at a glance. *Communications of the ACM*, 54(12):92–103, December 2011.
10. Maria Grazia Buscemi and Ugo Montanari. A survey of constraint-based programming paradigms. *Computer Science Review*, 2(3):137–141, December 2008.
11. Francesco Calimeri, Giovambattista Ianni, and Francesco Ricca. The third open answer set programming competition. *Theory and Practice of Logic Programming*, 14(1):117–135, January 2014.
12. Weidong Chen and David S. Warren. Tabled evaluation with delaying for general logic programs. *Journal of the ACM*, 43(1):20–74, January 1996.
13. Vašek Chvátal, William J. Cook, George B. Dantzig, Delbert R. Fulkerson, and Selmer M. Johnson. Solution of a large-scale traveling-salesman problem. In *50 Years of Integer Programming 1958-2008 - From the Early Years to the State-of-the-Art*, pages 7–28. Springer, 2010.
14. W. F. Clocksin and C. S. Mellish. *Programming in Prolog*. Springer, 1994.
15. Thomas H. Cormen, Charles E. Leiserson, Ronald L. Rivest, and Clifford Stein. *Introduction to Algorithms (3rd Edition)*. MIT Press, 2009.
16. Ron Cytron, Jeanne Ferrante, Barry K. Rosen, Mark N. Wegman, and F. Kenneth Zadeck. Efficiently computing static single assignment form and the control dependence graph. *ACM Transactions on Programming Languages and Systems*, 13(4):451–490, October 1991.

© The Author(s) 2015

N.-F. Zhou et al., *Constraint Solving and Planning with Picat*,

SpringerBriefs in Intelligent Systems, DOI 10.1007/978-3-319-25883-6

17. Brian Dean. Dynamic programming practice problems. http://people.cs.clemson.edu/~bcdean/dp_practice/.
18. Rina Dechter. *Constraint Processing*. Morgan Kaufmann, 2003.
19. Richard Fikes and Nils J. Nilsson. STRIPS: A new approach to the application of theorem proving to problem solving. *Artificial Intelligence*, 2(3–4):189–208, Winter 1971.
20. Martin Gebser, Roland Kaminski, Benjamin Kaufmann, and Torsten Schaub. *Answer Set Solving in Practice*. Morgan and Claypool, 2012.
21. Malik Ghallab, Dana S. Nau, and Paolo Traverso. *Automated Planning: Theory and Practice*. Morgan Kaufmann, 2004.
22. Hai-Feng Guo and Gopal Gupta. Simplifying dynamic programming via mode-directed tabling. *Software: Practice and Experience*, 38(1):75–94, January 2008.
23. Patrik Haslum and Ulrich Scholz. Domain knowledge in planning: Representation and use. In Enrico Giunchiglia, Nicola Muscettola, and Dana Nau, editors, *Proceedings of the International Conference on Automated Planning & Scheduling 2003 Workshop on PDDL*, pages 69–78, 2003.
24. Manuel V. Hermenegildo, Francisco Bueno, Manuel Carro, Pedro López-García, Edison Mera, José F. Morales, and Germán Puebla. An overview of Ciao and its design philosophy. *Theory and Practice of Logic Programming*, 12(1–2):219–252, January 2012.
25. Carl Hewitt. PLANNER: A language for proving theorems in robots. In *Proceedings of IJCAI'69 (International Joint Conferences on Artificial Intelligence)*, pages 295–302, 1969.
26. Paul Hudak, John Peterson, and Joseph H. Fasel. A gentle introduction to Haskell. https://www.haskell.org/tutorial/.
27. Tony Hürlimann. A coin puzzle: SVOR-contest 2007. www.svor.ch/competitions/competition2007/AsroContestSolution.pdf.
28. Joxan Jaffar and Michael J. Maher. Constraint logic programming: A survey. *The Journal of Logic Programming*, 19/20:503–581, May-July 1994.
29. Christopher Jefferson, Ian Miguel, Brahim Hnich, Toby Walsh, and Ian P. Gent. CSPLib: A problem library for constraints. http://www.csplib.org/, 1999.
30. Henry Kautz and Bart Selman. Planning as satisfiability. In Bernd Neumann, editor, *Proceedings of the 10th European Conference on Artificial Intelligence*, pages 359–363, 1992.
31. Henry Kautz and Bart Selman. The role of domain-specific knowledge in the planning as satisfiability framework. In Reid Simmons, Manuela Veloso, and Stephen Smith, editors, *Proceedings of the Fourth International Conference on Artificial Intelligence Planning Systems*, pages 181–189, 1998.
32. Richard E. Korf. Depth-first iterative-deepening: An optimal admissible tree search. *Artificial Intelligence*, 27(1):97–109, September 1985.
33. Robert Kowalski. *Logic for Problem Solving*. North Holland, Elsevier, 1979.
34. Steven M. LaValle. *Planning Algorithms*. Cambridge University Press, UK, 2006. Available at http://planning.cs.uiuc.edu/.
35. Vladimir Lifschitz. Answer set programming and plan generation. *Artificial Intelligence*, 138(1–2):39–54, June 2002.
36. John W. Lloyd. *Foundations of Logic Programming; (2nd Extended Edition)*. Springer, 1987.
37. Kim Marriott and Peter J. Stuckey. *Programming with Constraints: An Introduction*. MIT Press, 1998.
38. Kim Marriott, Peter J. Stuckey, Leslie De Koninck, and Horst Samulowitz. A MiniZinc tutorial. http://www.minizinc.org/downloads/doc-latest/minizinc-tute.pdf.
39. Drew McDermott. The planning domain definition language manual. CVC Report 98-003, Yale Computer Science Report 1165, 1998.
40. Mercury. http://www.mercurylang.org/.
41. Donald Michie. "Memo" functions and machine learning. *Nature*, 218(5136):19–22, April 1968.
42. Richard A. O'Keefe. *The Craft of Prolog*. MIT Press, 1994.

43. Gilles Pesant. A regular language membership constraint for finite sequences of variables. In Mark Wallace, editor, *Principles and Practice of Constraint Programming - CP 2004*, volume 3258 of *Lecture Notes in Computer Science*, pages 482–495. Springer, 2004. http://dx.doi.org/10.1007/978-3-540-30201-8_36.

44. Francesca Rossi. Constraint (logic) programming: A survey on research and applications. In Krzysztof R. Apt, Antonis C. Kakas, Eric Monfroy, and Francesca Rossi, editors, *New Trends in Contraints, Joint ERCIM/Compulog Net Workshop, Paphos, Cyprus, October 25-27, 1999, Selected Papers*, volume 1865 of *Lecture Notes in Computer Science*, pages 40–74. Springer, 1999. http://dx.doi.org/10.1007/3-540-44654-0_3.

45. Francesca Rossi, Peter van Beek, and Toby Walsh. *Handbook of Constraint Programming*. Elsevier, 2006.

46. Stuart Russell and Peter Norvig. *Artificial Intelligence: A Modern Approach (3rd Edition)*. Prentice Hall, 2009.

47. Konstantinos Sagonas and Terrance Swift. An abstract machine for tabled execution of fixed-order stratified logic programs. *ACM Transactions on Programming Languages and Systems*, 20(3):586–634, May 1998.

48. Vítor Santos Costa, Ricardo Rocha, and Luís Damas. The YAP Prolog system. *Theory and Practice of Logic Programming, Special Issue on Prolog Systems*, 12(1–2):5–34, January 2012.

49. Joachim Schimpf and Kish Shen. ECLiPSe—from LP to CLP. *Theory and Practice of Logic Programming*, 12(1–2):127–156, January 2012.

50. Jacob T. Schwartz, Robert B. K. Dewar, Ed Dubinsky, and Edith Schonberg. *Programming with Sets - An Introduction to SETL*. Springer, 1986.

51. Michael L. Scott. *Programming Language Pragmatics (3rd Edition)*. Morgan Kaufmann, 2009.

52. Robert W. Sebesta. *Concepts of Programming Languages (11th Edition)*. Pearson, 2015.

53. Leon Sterling and Ehud Shapiro. *The Art of Prolog*. MIT Press, 1997.

54. Hamdy A. Taha. *Operations Research: An Introduction (9th Edition)*. Prentice Hall, 2010.

55. Frank Takes. Sokoban: Reversed solving. In *Proceedings of the 2nd Nederlandse Studiev-ereniging Kunstmatige Intelligentie Student Conference*, pages 31–36. Citeseer, 2008.

56. Hisao Tamaki and Taisuke Sato. OLD resolution with tabulation. In Ehud Shapiro, editor, *Third International Conference on Logic Programming*, volume 225 of *Lecture Notes in Computer Science*, pages 84–98. Springer, 1986. http://dx.doi.org/10.1007/3-540-16492-8_66.

57. Edward Tsang. *Foundations of Constraint Satisfaction*. Academic Press, 1993.

58. Pascal van Hentenryck. *Constraint Satisfaction in Logic Programming*. MIT Press, 1989.

59. Willem-Jan van Hoeve and Irit Katriel. Global constraints. In Francesca Rossi, Peter van Beek, and Toby Walsh, editors, *Handbook of Constraint Programming*, Foundations of Artificial Intelligence, pages 169–208. Elsevier, 2006.

60. Toby Walsh. CSPLib problem 019: Magic squares and sequences. http://www.csplib.org/Problems/prob019.

61. Toby Walsh. CSPLib problem 024: Langford's number problem. http://www.csplib.org/Problems/prob024.

62. David H. D. Warren. WARPLAN: A system for generating plans. Technical Report DCL Memo 76, University of Edinburgh, 1974.

63. David H. D. Warren. An abstract Prolog instruction set. Technical note 309, SRI International, 1983.

64. David S. Warren. Memoing for logic programs. *Communications of the ACM*, 35(3):93–111, March 1992.

65. Wayne L. Winston. *Operations Research: Applications and Algorithms (4th Edition)*. Duxbury Press, 2003.

66. Neng-Fa Zhou. Parameter passing and control stack management in Prolog implementation revisited. *ACM Transactions on Programming Languages and Systems*, 18(6):752–779, November 1996.

67. Neng-Fa Zhou. Programming finite-domain constraint propagators in action rules. *Theory and Practice of Logic Programming*, 6(5):483–507, September 2006.

68. Neng-Fa Zhou. The language features and architecture of B-Prolog. *Theory and Practice of Logic Programming, Special Issue on Prolog Systems*, 12(1–2):189–218, January 2012.

69. Neng-Fa Zhou. Combinatorial search with Picat. *International Conference on Logic Programming, invited talk*, 2014. http://arxiv.org/pdf/1405.2538v1.pdf.

70. Neng-Fa Zhou, Roman Bartak, and Agostino Dovier. Planning as tabled logic programming. In *Theory and Practice of Logic Programming*, 2015.

71. Neng-Fa Zhou and Agostino Dovier. A tabled Prolog program for solving Sokoban. *Fundamenta Informaticae*, 124(4):561–575, 2013.

72. Neng-Fa Zhou and Jonathan Fruhman. A user's guide to Picat. http://picat-lang.org/download/picat_guide.pdf.

73. Neng-Fa Zhou and Jonathan Fruhman. Toward a dynamic programming solution for the 4-peg Tower of Hanoi problem with configurations. In Nicos Angelopoulos and Roberto Bagnara, editors, *Proceedings of CICLOPS 2012 (12th International Colloquium on Implementation of Constraint and Logic Programming Systems)*, pages 2–16, 2012. http://arxiv.org/pdf/1301.7673.pdf.

74. Neng-Fa Zhou and Christian Theil Have. Efficient tabling of structured data with enhanced hash-consing. *Theory and Practice of Logic Programming*, 12(4–5):547–563, July 2012.

75. Neng-Fa Zhou, Yoshitaka Kameya, and Taisuke Sato. Mode-directed tabling for dynamic programming, machine learning, and constraint solving. In *22nd IEEE International Conference on Tools with Artificial Intelligence*, pages 213–218, 2010.

76. Neng-Fa Zhou, Taisuke Sato, and Yi-Dong Shen. Linear tabling strategies and optimizations. *Theory and Practice of Logic Programming*, 8(1):81–109, January 2008.

Index

! =, 9
! ==, 9
++, 6, 7
: =, 12, 23
==, 9, 23
=, 9, 23
#=/2, 35
#>/2, 35
action/4, 102
all_different/1, 33, 35, 36, 46, 63
all_different_except_0/1, 58
append/3, 23, 26
apply/n, 23
backward, 62
best_plan/4, 116
best_plan_bb/4, 116
best_plan_unbounded/4, 102, 103
call/n, 23
circuit/1, 70, 131
cl/1, 2
cond, 12
constr, 62
cumulative/4, 65, 67
current_resource/0, 116
degree, 62
down, 62
element/3, 131
fail/0, 80
ffc, 62
ffd, 62
ff, 62
final/1, 102
find_all, 80
fold/3, 23
forward, 62
freeze/2, 23
get/2, 8

get/3, 8
global_cardinality/2, 68, 69
ground/1, 3
has_key/2, 8
import, 10, 35
index, 18
initialize_table/0, 91
inout, 62
leftmost, 62
len/1, 24
length/1, 6, 7, 29
lex/2, 57
lex2/1, 57, 81
lex_le/2, 57
map/2, 23
max/1, 17
max, 62
membchk/2, 21, 26
member/2, 2, 17, 21, 26
min/1, 17
min, 62
new_array/1, 7
new_list/1, 6
new_map/1, 8
new_map/2, 8
nl, 10
once, 11, 26
plan/4, 116, 117
plan_unbounded/4, 102, 103, 117
print/1, 10
println/1, 10
put/3, 8
rand_val, 62
rand_var, 62
read_int/1, 10
regular/6, 74, 75, 80, 81
remove_dups/1, 22

© The Author(s) 2015
N.-F. Zhou et al., *Constraint Solving and Planning with Picat*,
SpringerBriefs in Intelligent Systems, DOI 10.1007/978-3-319-25883-6